Engineering Chemistry

Engineering Chemistry

(*Third Edition*)

Dr. R.V. Gadag
Professor (Retd.)

Dr. A. Nityananda Shetty
Professor
Department of Chemistry
National Institute of Technology
Karnataka, Shurathkal

TECHSAR™
NEW DELHI

Published and Printed by

Techsar Pvt. Ltd.
4435-36/7, Ansari Road, Daryaganj,
New Delhi - 110 002
E-mail: tech_info@techsarworld.com
Website: www.techsarworld.com

ISBN 978-93-84588-02-1

Reprint 2025

Preface to the Third Edition

We present the third edition of ***Engineering Chemistry***, with a few changes in the contents of the book. The style of the book has been retained to be lucid, as in the previous editions. The topics have been dealt with in-depth explanations using simple and clear language.

We are thankful to all the students and fellow teachers for their overwhelming response to the earlier editions of the book. Their valuable response and inputs have immensely helped us to improve the presentation and contents in the present edition.

We hope that the present edition of the book will meet the expectations of the readers. Feedback from the readers, for the further improvement of the book will be gratefully accepted.

R. V. Gadag
A. Nityananda Shetty
Email: nityashreya@gmail.com

Preface to the First Edition

he recent advances in the field of engineering and technology have brought phenomenal changes in ıman lifestyle, improving the personal comforts in terms of housing, communication, transportation, althcare, etc. Technological revolutions have made the facilities affordable even to common man. chnologies responsible for enhancing the quality of human life are evolved on scientific principles. nerefore, it is demanding from a successful engineer to be well versed with the knowledge of basic sciences like chemistry and physics. This book, *Engineering Chemistry*, has been written keeping hese facts in consideration to fulfil the objectives of engineering education.

The book comprises a few chapters which deal with the basic principles of chemistry and a few her chapters on applied aspects of chemistry. The topics covered are expected to provide an interphase etween principles of chemistry and engineering. The contents of the book are directed at providing o the engineering students, sufficient insight into the chemistry principles.

The book has been written in a simple language with in depth explanations. The book contains suitable neat diagrams and solved numerical problems wherever required. The review questions at the end of each chapter will help the students in revising the topics that they have studied. It is the fond hope of the authors that the book will be highly useful to both the student community and the teachers.

The authors are highly optimistic of receiving encouraging response from the readers, similar to the one they received for their earlier books. Any constructive comments and suggestions from our valued readers for the further improvement of the book will be gratefully accepted.

R.V. Gadag
A. Nityananda Shetty

Contents

Laboratory Manual in Engineering Chemistry

CHAPTER 1

Electrochemical Cells

1.1. INTRODUCTION

Electrochemistry is a branch of science, which studies conversion of chemical energy into electrical energy and vice versa. The postulates that chemical and electrical phenomenon are mutually related were put forward by M. Lomonsov as early as in the middle of the 18th century. However, researchers drew their attention to a new field only after L. Galvani reported his investigations of the contraction of muscles in dissected frogs when these muscles were connected to copper and iron wires. In 1800, A. Volta made the first source of electrical energy, a voltaic pile. In 1872, F. D. Engels suggested that chemical reactions taking place in galvanic cells were the sources of electrical energy. The basis for the theory of electrolysis was laid down by T. Grotgus in 1805. The essence of galvanic cell energy was elucidated by D. Gibbs and H. Helmholtz in 1882 and in 1889. V. Nernst derived an equation relating emf of galvanic cell to the solution concentration. The theoretical concepts were developed by several scientists and some of them include I. Hittorf (1853), F. Kohlrausch (1874), S. Arrhenius (1887), P. Debye and L. Huckel (1923).

The development of theoretical concepts of electrochemistry was accompanied by practical applications of electrochemical phenomenon. Electrolysis phenomenon opened up new spheres of applications such as electroplating, electro-extraction of metals, electro-refining of metals, electro-production of certain compounds, etc. Various types of galvanic cells serve as very effective portable energy sources. Electrochemical principles helped in understanding the corrosion of metal and in combating it efficiently.

Because of the practical importance of electrochemistry, it has become a very interesting field of study for the scientists and technologists. This chapter deals with a few fundamental concepts related with electrochemical cells.

1.2. ELECTROCHEMICAL CELLS

An **electrochemical cell**is a device which converts chemical energy into electrical energy or electrical energy into chemical energy. Thus, there can be two types of electrochemical cells.

(i) The one which converts chemical energy into electrical energy – the so called **galvanic cell** or **voltaic cell**.

(ii) The one which converts electrical energy into chemical energy – the **electrolytic cell.**

1.3. GALVANIC CELLS

Galvanic cells are devices which convert chemical energy into electrical energy. Galvanic cell generally consists of two electrodes dipped in two electrolyte solutions which are separated by a porous diaphragm or connected through a salt bridge. To illustrate a typical galvanic cell, we can take the example of Daniell cell.

The Daniell cell consists of two containers, one of which is with a zinc rod dipped in zinc sulfate solution and the other is with a copper rod dipped in copper sulfate solution. The solutions of the two containers are connected by a salt bridge. The salt bridge is a U-tube, filled with either a jelly containing KCl or NH_4NO_3, or a saturated solution of KCl or NH_4NO_3 and both the ends of the U-tube plugged with a porous material. When the zinc and copper rods (electrodes) are connected externally through an ammeter by using a wire, the following changes are observed:

- Zinc starts to dissolve
- Copper gets deposited on copper rod
- The ammeter indicates the flow of electrons from zinc to copper rod
- The zinc sulfate solution becomes richer with Zn^{2+} ions
- The copper sulfate solution become more dilute with respect to Cu^{2+} ions
- Migration of ions takes place in the solution.

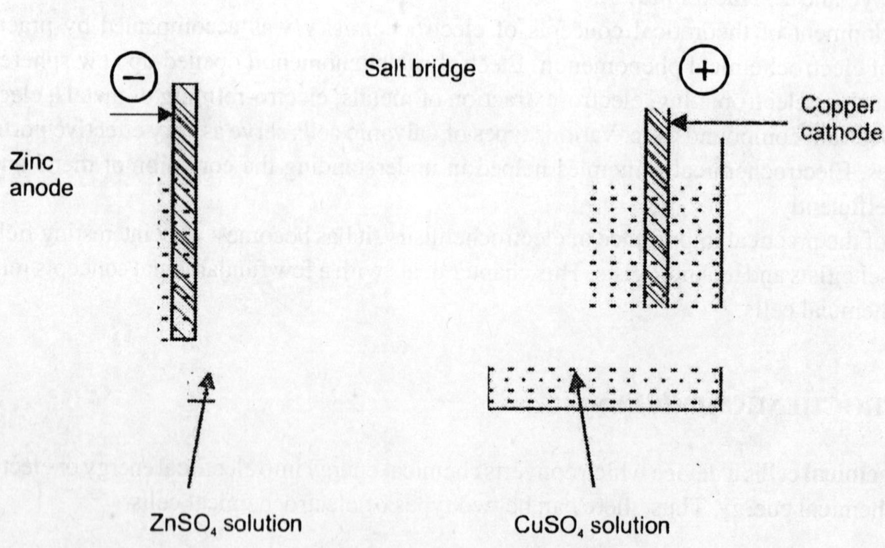

Fig. 1.1. Daniell cell

These processes will continue as long as zinc rod in zinc sulfate solution and Cu^{2+} ions in copper sulfate solution remain and connections are maintained.

The above observations are explained by the following oxidation-reduction reactions taking place at the electrodes.

At the zinc electrode, zinc goes into the solution as Zn^{2+} ions, liberating two electrons.

$$Zn \rightarrow Zn^{2+} + 2e^-$$

In the process zinc undergoes oxidation or in other words at the zinc electrode oxidation takes place.

On the other hand, at the copper electrode reduction of Cu^{2+} ions takes place with the deposition of metallic copper on copper rod, by consuming two electrons.

$$Cu^{2+} + 2e^- \rightarrow Cu$$

Thus electrons are liberated at the zinc electrode, and electrons are consumed at the copper electrode. Therefore, there is an excess of electrons at the zinc electrode and dearth of the same at the copper electrode. When the circuit is closed, electrons flow from zinc to copper through the external wire to feed the Cu^{2+} ions in the process of discharging copper. Thus electrical energy or current is produced which may be used for the purpose of external work.

The net chemical change taking place is,

$$Zn \rightarrow Zn^{2+} + 2e^- \quad \text{(Oxidation)}$$

$$Cu^{2+} + 2e^- \rightarrow Cu \quad \text{(Reduction)}$$

$$Zn + Cu^{2+} \rightarrow Zn^{2+} + Cu \quad \text{(Cell reaction)}$$

Thus the overall reaction taking place in the cell is the redox reaction.

By convention, the electrode at which oxidation occurs is **anode** and the electrode where reduction occurs is **cathode**. Thus in the above cell zinc electrode acts as anode and copper acts as cathode. In a galvanic cell anode is **negative** and cathode is **positive**. The reason, to this nomenclature is that oxidation is accompanied by the liberation of electrons which are given up to that electrode, which thereby acquires a negative charge. Reduction, on the other hand, is accompanied by the consumption of electrons by the reactant in the solution from the electrode, which thereby acquires a positive charge.

Hence the flow of electrons is from the zinc electrode to the copper electrode. The direction of flow of positive electricity, which is taken as the conventional current, is from the copper to the zinc electrode.

The net cell reaction, called the **electrochemical reaction** is the cause for the production of electricity. The net reaction is the sum of the two electrode reactions, of which one is the oxidation, the other is reduction. Each electrode reaction is called as **half cell reaction**.

As in the case of Daniell cell, the combination of any two different electrodes constitutes a galvanic cell. The one with greater tendency to go into the solution by liberating electrons, acts as the anode, and the other one with more tendency to get deposited acts as the cathode.

1.4. ELECTROMOTIVE FORCE OF THE CELL

Current can not flow from one point to another unless there is a potential difference between the two points. Hence, a flow of electrons from anode to cathode in a galvanic cell indicates that the two electrodes have different potentials. "The potential difference between the two electrodes of a galvanic cell which causes the flow of current from one electrode (higher potential) to the other (lower potential) is called the **electromotive force (emf)** of the cell or the **cell potential**".

The emf of a cell is denoted by E_{cell}. The emf of a cell depends on the nature of the electrode, temperature and concentration of the electrolyte solutions. To facilitate comparison of different galvanic cells, their **standard emf** values, denoted by E°_{cell} are used.

Standard emf of a cell is defined as the emf of a galvanic cell when the reactants and products of the cell reaction are at unit concentration or unit activity, at 298 K and at 1 atmospheric pressure.

The emf of a cell represents the driving force of a cell reaction. As represented by the thermodynamic relation,

$$\Delta G = -\, nFE \qquad (1.1)$$

where ΔG is the free energy change accompanying the cell reaction, n is the number of electron transfer during the cell reaction and F is faraday.

The cell reaction is spontaneous or feasible when ΔG is negative. ΔG can be negative only if emf of the cell, E is positive, because, the other two factors, n and F are always positive. Thus emf of a galvanic cell is always positive. The positive emf value indicates the spontaneity of the cell reaction in the given direction.

Measurement of emf of a cell

The emf of a cell can not be measured accurately by using a voltmeter. This is because, in the process of measuring the emf by using a voltmeter, some current is drawn from the cell that alters the concentration of the electrolyte and thereby causing a change in the emf of the cell. And also a part of the emf is used to overcome the internal resistance of the cell.

Potentiometric method

Principle: The emf of a cell can be measured accurately by potentiometric method, which is based on Poggendorffs compensation principle. In this method, the emf of a test cell is opposed by the impressed emf from an external source of emf, and the measurements are made when there is no net flow of current in the circuit.

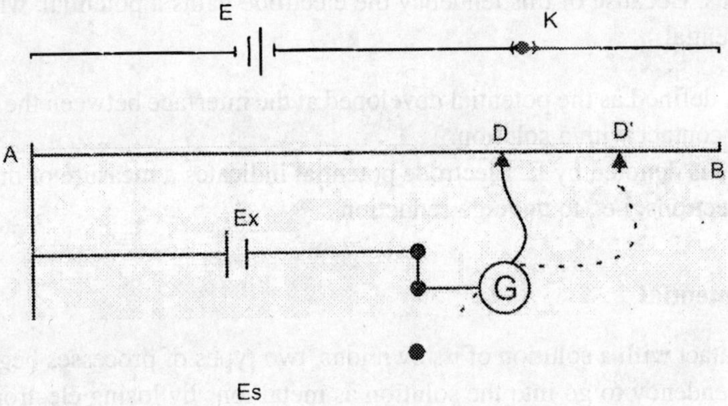

Fig. 1.2. Measurement of emf of a cell

Procedure: A simplified diagram of potentiometer is shown in Fig. 1.2. The two terminals of a storage battery (E) which has a constant and higher emf than the test cell, is connected to two ends of a long uniform wire AB of high resistance. The positive terminal of the test cell is connected to the end A, to which the positive terminal of E is connected and the negative terminal is connected to a sliding contact D through a galvanometer (G). The sliding contact is moved along the wire AB until the null point is reached, which is indicated by no current flow in the galvanometer. The position of null point is noted and the distance AD is measured. The emf of the cell E_x is proportional to the length AD.

$$E_x \alpha AD$$

Then the test cell E_x is replaced by a standard cell E_s, whose emf is known. The position of the sliding contact is readjusted by moving it over AB, as before, till the null point D^1 is reached again. The length AD^1 is measured. The emf of the standard cell E_s is proportional to the length AD^1.

$$E_x \alpha AD^1$$

Therefore, $$E_x / E_s = AD/AD^1$$

And, $$E_x = (AD/AD^1) \times E_s$$

Thus the emf of the test cell is calculated by knowing the emf of the standard cell.

A **standard cell** is the one which has a constant and reproducible known emf with a negligible variation of emf with temperature. The most commonly used standard cell is **Weston standard cell.**

1.5. SINGLE ELECTRODE POTENTIAL

A galvanic cell essentially consists of two electrodes or half cells. At one electrode oxidation reaction takes place, i.e., electrons are liberated. At the other electrode, reduction reaction takes place, i.e., electrons are consumed. Whenever, a metal is in contact with its own ions, it has a natural tendency to lose or gain electrons. Because of this tendency the electrode gains a potential, which is known as single electrode potential.

Electrode potential is defined as the potential developed at the interface between the metal and the solution, when it is in contact with a solution.

Electrode potential is denoted by E. Electrode potential indicates a measure of the tendency of an electrode to gain electrons, i.e., to undergo reduction.

Origin of electrode potential

When a metal is in contact with a solution of its own ions, two types of processes begin to operate. The metal shows the tendency to go into the solution as metal ions by losing electrons.

$$M \rightarrow M^{n+} + ne^-$$

At the same time, metal ions in the solution show the tendency to get deposited as metal atoms.

$$M^{n+} + ne^- \rightarrow M$$

Depending upon the nature of the metal, either the dissolution of metal or the deposition of metal is faster in the beginning and the other one is slower. But as the reactions proceed, the faster reaction becomes slower and the slower reaction becomes faster and a state of equilibrium is reached after some time.

$$M^{n+} + ne^- \rightleftharpoons M$$

Further, the number of metal atoms dissolving is equal to the number of atoms getting deposited. But the net change at the equilibrium is dependent on the nature of the metal.

If the dissolution reaction is faster than the deposition reaction, the net change when equilibrium is attained, is a few metal atoms have dissolved in the solution, liberating electrons. These electrons accumulate on the electrode surface, making it negatively charged (Fig. 1.3.(a)).

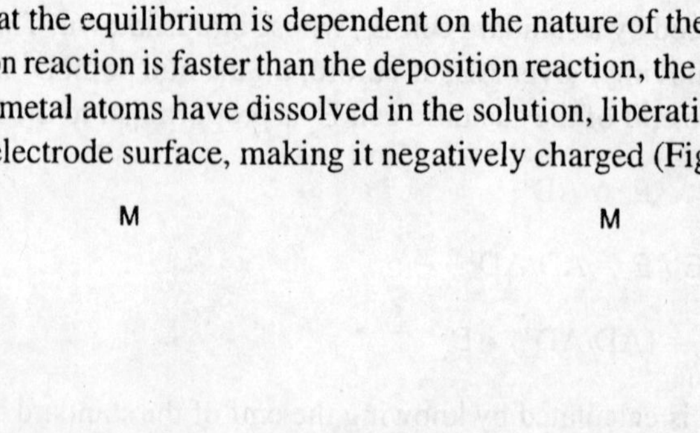

Fig. 1.3

The negatively charged electrode surface attracts a layer of positively charged ions at the interface, developing an electrical double layer at the metal-solution interface. Similarly, if the deposition reaction is faster than dissolution, at equilibrium, the net change is deposition of few metal ions as metal atoms by consuming electrons. As a result the electrode surface develops a layer of positive charges which attracts a layer of negatively charged ions at the interface (Fig. 1.3 (b)), again establishing an electrical double layer. Formation of an electrical double layer gives rise to a potential difference across the layer. This potential difference between the metal and the solution at the interface is the single electrode potential.

Hydrogen scale of electrode potential

It is not possible to determine the single electrode potential experimentally. Only the difference in electrode potentials between any two electrodes can be measured as emf of the cell formed by the combination of the electrodes. In order to have an operative value for electrode potential, electrode potential values are assigned to different electrodes with respect to the standard hydrogen electrode. The electrode potential of the standard hydrogen electrode is arbitrarily fixed at zero and the numerical values of the electrode potentials of various electrodes are relative values with respect to standard hydrogen electrode.

Therefore, single electrode potentials of electrodes are referred to as potentials on the **hydrogen scale**.

According to the latest accepted conventions, all single electrode potential values represent reduction tendency of electrodes. Higher the value of electrode potential, higher is the tendency of the electrode to act as a reduction electrode.

When two electrodes are combined to form a cell, the one with lower electrode potential, with lower tendency to undergo reduction acts as anode. The one with higher electrode potential, with higher tendency to undergo reduction acts as cathode. For example, in Daniell cell, zinc with lower electrode potential acts as anode and copper with higher electrode potential acts as cathode. Similarly, when copper electrode is combined with a silver electrode, copper with lower electrode potential, acts as anode and silver with higher electrode potential, acts as cathode.

Standard electrode potential

Standard electrode potential is the electrode potential when the electrode is in contact with a solution of unit concentration at 298 K. If the electrode involves a gas, then the gas is at 1 atmospheric pressure. It is denoted by E°.

Sign of electrode potential

The electrode potential of an electrode is positive, if the electrode reaction is reduction when coupled with the standard hydrogen electrode and is negative if the electrode reaction is oxidation when coupled with the standard hydrogen electrode.

For example, when copper electrode is combined with standard hydrogen electrode, copper electrode acts as cathode and undergoes reduction and hydrogen electrode acts as anode.

$$H_2\,(g) \rightarrow 2H^+ + 2e^- \text{ (oxidation)}$$
$$Cu^{2+} + 2e^- \rightarrow Cu \text{ (reduction)}$$

Hence electrode potential of copper is assigned a positive sign. Its standard electrode potential is 0.34 V.

When zinc is coupled with standard hydrogen electrode, zinc electrode acts as the anode and hydrogen electrode acts as the cathode.

$$Zn \rightarrow Zn^{2+} + 2e^-$$
$$2H^+ + 2e^- \rightarrow H_2$$

Hence, the electrode potential of zinc is negative.

A standard hydrogen electrode is one in which the hydrogen gas at 1 atmospheric pressure is bubbled through a solution of hydrogen ions of unit activity or unit concentration at 298 K.

1.6. NERNST EQUATION

In 1889, Nernst derived a quantitative relationship between electrode potential and concentration of the electrolyte species involved. This expression is known as **Nernst equation.**

Consider the following reversible electrode reaction:

$$M^{n+} + ne^- \rightleftharpoons M$$

By convention, the electrode reaction is written with reduction in the forward direction, as electrode potential is the measure of reduction tendency.

The decrease in free energy, ΔG, accompanying the process is given by the well known thermodynamic equation,

$$-\Delta G = -\Delta G^\circ - RT\ln K \quad (1.2)$$

where ΔG° is the decrease in free energy accompanying the same process when the reactants and products of the reaction are in their standard states of unit activity, K stands for reaction quotient of the activities of the products and reactants at any given stage of the reaction.

Substituting the value of K, we have,

$$-\Delta G = -\Delta G^\circ - RT\ln \quad (1.3)$$

As an approximation, substituting molar concentration for activities,

$$-\Delta G = -\Delta G^\circ - RT\ln \frac{[M]}{[M^{n+}]} \quad (1.4)$$

If E is the electrode potential of the electrode in volts, and the electrode reaction involves transfer of n electrons, i.e., nF coulombs, the electrical work available from the electrode is nFE volt coulomb or joules. Hence free energy decrease of the system ΔG is given by the expression,

$$-\Delta G = nFE \tag{1.5}$$

Hence, $$nFE = -\Delta G^\circ - RT\ln\frac{[M]}{[M^{n+}]}$$

$$= nFE^\circ - RT\ln\frac{[M]}{[M^{n+}]} \quad (\text{as } -\Delta G^\circ = -\,nFE^\circ)$$

or $$E = E^\circ - \frac{RT}{nF}\ln\frac{[M]}{[M^{n+}]}$$

$$= E^\circ + \frac{RT}{nF}\ln[M^{n+}] \quad (\text{Since } [M] = 1)$$

$$E = E^\circ + \frac{2.303\,RT}{nF}\log[M^{n+}] \tag{1.6}$$

where E° is the standard electrode potential, R is the gas constant, T is the absolute temperature and F is Faraday.

The equation 1.6 is the **Nernst equation** for electrode potential. The equation can be used to calculate the electrode potential of an electrode if the concentration is known. At 298K, when the values of R,T and F are substituted, the equation 1.6 reduces to,

$$E = E^\circ + \frac{0.0591}{n}\log[M^{n+}] \text{ at } 298\text{ K} \tag{1.7}$$

The Nernst equation can also be applied for the calculation of emf of a cell. Consider the cell reaction,

$$aA + bB \rightleftharpoons cC + dD$$

The Nernst equation for the emf of the cell is

$$E_{cell} = E^\circ_{cell} - \frac{2.303\,RT}{nF}\log\frac{[C]^c[D]^d}{[A]^a[B]^b} \tag{1.8}$$

$$E_{cell} = E^\circ_{cell} - \frac{0.0591}{n}\log\frac{[C]^c[D]^d}{[A]^a[B]^b} \text{ at } 298\text{ K} \tag{1.9}$$

where n is the number of electrons transferred during the cell reaction and E°_{cell} is the standard emf of the cell.

Example 1.1. Calculate the electrode potential of a zinc electrode dipped in a solution of 0.05M $ZnSO_4$ in which the degree of dissociation of $ZnSO_4$ is 0.6 at 298 K. Standard electrode potential of zinc electrode is – 0.76V.

Solution

The electrode reaction is, $Zn^{2+} + 2e^- \rightleftharpoons Zn$,

Concentration of $ZnSO_4$ solution 0.05M

Degree of dissociation of $ZnSO_4$ 0.6

Therefore, Concentration of Zn^{2+} ions, $[Zn^{2+}]$ 0.05 × 0.6

0.03 M

Standard electrode potential, E° – 0.76 V

Temperature, T 298 K

n 2

∴ At 298 K, Electrode potential,

$$E \quad E^\circ + \frac{0.0591}{n} \log [Zn^{2+}]$$

$$-0.76 + \frac{0.0591}{2} \log 0.03$$

$$\mathbf{-0.805\ V}$$

Example 1.2. Calculate the emf of a cadmium – copper cell in which cadmium is in contact with 0.002M $CdSO_4$ and copper in contact with 0.02 M $CuSO_4$ solution. The standard emf of the cell is 0.74 V at 298 K.

Solution:

The cell reaction is,

$$Cd + Cu^{2+} \rightleftharpoons Cd^{2+} + Cu$$

$[Cu^{2+}]$ 0.02 M

$[Cd^{2+}]$ 0.002 M

E°_{cell} 0.74 V

n 2

At 298 K, E.M.F of the cell, E

$$E^\circ_{cell} - \frac{0.0591}{n} \log \frac{[Cd^{2+}][Cu]}{[Cd][Cu^{2+}]}$$

$$E^\circ_{cell} - \frac{0.0591}{n} \log \frac{[Cd^{2+}]}{[Cu^{2+}]} \quad (\text{as } [Cu] = [Cd] = 1)$$

$$= 0.74 - \frac{0.0591}{2} \log \frac{0.002}{0.02}$$

$$= 0.74 - 0.02955 \log 0.1$$

$$= 0.74 + 0.02955$$

$$= \mathbf{0.7695\ V}$$

1.7. ELECTROCHEMICAL CONVENTIONS

(i) Sign of the electrode potential: The electrode potential of an electrode is positive if reduction is the electrode reaction, when it is combined with a standard hydrogen electrode. The electrode potential is negative if the electrode reaction is oxidation when the electrode is combined with a standard hydrogen electrode.

(ii) Representation of an electrode: If an electrode acts as anode, i.e. if oxidation takes place at the electrode, then the electrode is represented as $\mathbf{M_1/M_1^{n+}\ (C_1)}$.

Here the convention is that, we have metal atoms first which later gets converted into metal ions as given by the equation,

$$M \rightarrow M^{n+} + ne^-$$

The vertical line represents the fact that the metal atoms and ions are in different phases. The term inside the bracket (C_1) represents the concentration of the metal ion. If reduction takes place at the electrode, then the reaction is

$$M_2^{n+} + ne^- \rightarrow M_2$$

Using the above convention, this electrode can be represented as $\mathbf{M_2^{n+}(C_2) / M_2}$. In this case, we have the ions first and then on reduction they get converted into atoms.

(iii) Representation of a cell: If two electrodes are combined to form a cell, then the cell is represented by writing the electrode conventions side by side with the anode (negative electrode) on the left and the cathode (positive electrode) on the right. The two electrode representations are separated by a double line. The double line represents the presence of a sait bridge. Hence, the cell can be represented as:

$$\mathbf{M_1/M_1^{n+}(C_1)//M_2^{n+}(C_2)/M}$$

e.g.: Daniell cell can be represented as follows:

$$Zn/Zn^{2+}(1M)//Cu^{2+}(1M)/Cu$$

(iv) Calculation of cell emf:

The cell emf, E_{cell} $E_{rhse} - E_{lhse}$

$E_{cathode} - E_{anode}$

Standard emf, $E^\circ_{cell} = E^\circ_{cathode} - E^\circ_{anode}$

Therefore, by knowing the electrode potentials of the electrodes constituting the cell, the emf of the cell can be calculated.

(v) Writing electrode reactions and cell reaction: The cell reaction is the sum of the electrode reactions as they occur in the cell. At the anode oxidation takes place and at the cathode reduction is the electrode reaction.

For example, in Daniell cell,

$$Zn \rightarrow Zn^{2+} + 2e^- \text{ (Anode)}$$

$$Cu^{2+} + 2e^- \rightarrow Cu \text{ (Cathode)}$$

$$Zn + Cu^{2+} \rightarrow Zn^{2+} + Cu \text{ (Cell reaction)}$$

The cell reaction is spontaneous and is so only when emf of the cell is positive. This can be understood from the relation, $\Delta G = -nFE$.

For the spontaneity of a reaction, ΔG should be negative, which is possible only when E is positive as n and F are always positive.

If the wrong assumption be made with respect to the polarity of the electrodes, the calculation of emf from electrode potential values will yield a negative emf, corresponding to a non-spontaneous reaction. In such cases, the polarity of the electrodes should be reversed and the cell reaction actually occurs spontaneously in the reverse direction.

Function of salt bridge

In a galvanic cell, if the salt bridge is not used, then the two solutions should be either in contact with each other or separated by a porous membrane. In either case, the positive and negative ions of electrolytes will migrate across the junction in opposite directions as there is increase of positive and negative charges at anode and cathode respectively. Inside the cell the current is carried by the ions, negative ions from cathode to anode and positive ions from anode to cathode. Because of the difference in the migration velocities of these ions, accumulation of charges across the junction occurs. This gives rise to the formation of an electrical double layer developing a potential called liquid junction potential across the junction. In such situations, the emf of the cell includes liquid junction potential also.

The use of salt bridge, however, avoids the liquid junction potential and provides the contact between the two solutions. It consists of a saturated solution of a salt such as KCl or NH_4NO_3, whose ions have almost same migration velocities. The positive and negative ions of the salt in the salt bridge migrate with equal speed into cathode and anode compartments respectively, there by avoiding liquid junction potential.

The salt bridge is prepared by dissolving a little of agar agar in a hot solution of KCl and filling in a U-tube. On cooling, the agar forms a gel with KCl solution in it. The salt bridge is also prepared by filling a U-tube with saturated solution of KCl and plugging the two ends with a porous material.

Example 1.3. Give the representation of a standard cadmium – copper cell, write the electrode and cell reactions and calculate the standard emf of the cell, if standard electrode potentials of cadmium and copper are – 0.4 and 0.34 V respectively.

Solution:

Since $E^\circ_{Cd} < E^\circ_{Cu}$, cadmium acts as anode and copper acts as cathode.

Cell representation

$Cd / Cd^{2+}(1M) // Cu^{2+}(1M) / Cu$

Electrode reactions

At the anode, $Cd \rightarrow Cd^{2+} + 2e^-$

At the cathode, $Cu^{2+} + 2e^- \rightarrow Cu$

Cell reaction, $Cd + Cu^{2+} \rightarrow Cd^{2+} + Cu$

Standard e.m.f. E°_{cell} $E^\circ_{cathode} - E^\circ_{anode}$

$E^\circ_{Cu} - E^\circ_{Cd}$

$0.34 - (-0.40)$

0.74 V

Example 1.4. Write the cell representation, cell reaction and calculate the standard emf of a Ni – Ag cell if E°_{Ni} and E°_{Ag} are – 0.25 and 0.8V, respectively.

Solution:

Cell representation: $Ni / Ni^{2+}(1M) // Ag^+(1M) / Ag$

Cell reaction:

At the anode, $Ni \rightarrow Ni^{2+} + 2e^-$

At the cathode, $2Ag^+ + 2e^- \rightarrow 2Ag$

Cell reaction, $Ni + 2Ag^+ \rightarrow Ni^{2+} + 2Ag$

Standard e.m.f. of the cell, E°_{cell} $E^\circ_{Ag} - E^\circ_{Ni}$

$0.8 - (-0.25)$

$= \mathbf{1.05V}$

Example 1.5. For the cell, $Fe / Fe^{2+}(0.01) // Ag^{+1}(0.1) / Ag$ write the cell reaction and calculate the emf of the cell at 298 K, if standard electrode potentials of Fe and Ag electrodes are –0.44 and 0.8V respectively.

Solution:

Cell reaction: At the anode, $Fe \rightarrow Fe^{2+} + 2e^-$

At the cathode, $2Ag^+ + 2e^- \rightarrow 2Ag$

Cell reaction, $Fe + 2Ag^+ \rightarrow Fe^{2+} + 2Ag$

E°_{cell} $E^\circ_{Ag} - E^\circ_{Fe}$

$0.8 - (-0.44)$

$1.24V$

By applying Nernst equation,

$$E_{cell} = E^\circ_{cell} - \frac{0.0591}{n} \log \frac{[Fe^{2+}][Ag]^2}{[Fe][Ag^+]^2}$$

$$E^\circ_{cell} - \frac{0.0591}{n} \log \frac{[Fe^{2+}]}{[Ag^+]^2}, \text{ (as [Fe] = [Ag] = 1)}$$

$$1.24 - \frac{0.0591}{2} \log \frac{0.01}{0.1^2}$$

$1.24 - 0$

$\mathbf{1.24\ V}$

Exercise 1.6. The e.m.f of a cell, $Mg / Mg^{2+} (0.01) // Cu^{2+}(C_1) / Cu$ is measured to be 2.78V at 298K. The standard electrode potential of magnesium electrode is – 2.371V. Calculate the electrode potential of copper electrode.

Solution

Cell reaction: At the anode, $Mg \rightarrow Mg^{2+} + 2e^-$

At the cathode, $Cu^{2+} + 2e^- \rightarrow Cu$

Cell reaction, $Mg + Cu^{2+} \rightarrow Mg^{2+} + Cu$

Electrode potential of Mg electrode,

$$E_{Mg} = E^\circ_{Mg} + \frac{0.0591}{n} \log [Mg^{2+}]$$

$$= -2.37 + \frac{0.0591}{2} \log 0.01$$

$$= -2.37 - 0.0591$$

$$= \mathbf{-2.4291V}$$

$$E_{cell} = E_{Cu} - E_{Mg}$$

$$2.78 = E_{Cu} - (-2.4291)$$

$$E_{Cu} = 2.78 - 2.4291$$

$$= \mathbf{0.3509V}$$

1.8. ELECTROCHEMICAL SERIES

The arrangement of elements in the order of their standard electrode potential is referred to as emf or electrochemical series. Such an arrangement of a few elements is given in Table 1.1.

Table 1.1 Electrochemical Series

M^{n+} / M	E° (V) at 298K	M^{n+} / M	E° (v) at 298K
Li^+ / Li	− 3.05	Cd^{2+} / Cd	−0.40
K^+ / K	− 2.93	Sn^{2+} / Sn	− 0.14
Ba^{2+} / Ba	− 2.90	Pb^{2+} / Pb	− 0.13
Ca^{2+} / Ca	− 2.87	H^+ /H2	0.00
Na^+ / Na	− 2.71	Cu^{2+} / Cu	0.34
Mg^{2+} / Mg	− 2.37	Ag^+ / Ag	0.80
Al^{3+} / Al	− 1.66	Hg^{2+} / Hg	0.85
Zn^{2+} / Zn	− 0.76	Pt^{2+} / Pt	1.20
Fe^{2+} / Fe	− 0.44	Au^{3+} / Au	1.38

- A negative value indicates oxidation tendency and a positive value indicates reduction tendency with respect to hydrogen.
- The metals with lower electrode potential values are more reactive and as the electrode potential increases, the reactivity decreases and the metals with higher electrode potentials are more noble.
- Metals with lower electrode potentials have the tendency to replace metals with higher electrode potential from their solutions. For example, Zn displaces Cu, Cu displaces Ag.
- Metals with negative electrode potentials can liberate hydrogen from acidic solutions.

1.9 TYPES OF ELECTRODES

In electrochemical work, the cells encountered involve various types of electrodes depending upon the purpose for which they are used. Some of the important types may be grouped into the following types:

(i) Metal-metal ion electrode: An electrode of this type consists of a metal dipped in a solution containing its own ions. e.g., Zn / Zn^{2+}, Cu / Cu^{2+}, Ag / Ag^{+}.

The electrode reaction is, in general,

$$M^{n+} + ne^- \rightleftharpoons M$$

(ii) **Metal – metal salt ion electrode (Metal – insoluble salt electrode):** These electrodes consist of a metal in contact with a sparingly soluble salt of the same metal dipped in a solution containing anion of the salt.

e.g., Calomel electrode ($Hg / Hg_2Cl_2 / Cl^-$)

Silver – Silver chloride electrode ($Ag / AgCl / Cl^-$)

Lead – lead sulphate electrode ($Pb / PbSO_4 / SO_4^{2-}$)

(iii) Gas electrode: Gas electrodes consist of a gas bubbling about an inert metal wire or foil, immersed in a solution containing ions to which the gas is reversible. The metal provides electrical contact and facilitates the establishment of equilibrium between the gas and its ions.

e.g., H_2 electrode ($Pt / H_2 / H^+$)

Chlorine electrode ($Pt / Cl_2 / Cl^-$)

(iv) Amalgam electrode: Amalgam electrode is similar to metal-metal ion electrode except for the fact that in an amalgam electrode, metal-amalgam is in contact with a solution containing its own metal ions.

e.g., Lead amalgam electrode ($Pb - Hg / Pb^{2+}$)

(v) Oxidation-reduction electrode: An oxidation-reduction electrode is the one in which the electrode potential arises from the presence of oxidised and reduced forms of the same substance in solution.

The potential arises from the tendency of one form to change into the other more stable form. The potential developed is picked up by an inert electrode like platinum.

e.g., $Pt / Fe^{2+}, Fe^{3+}$

$Pt / Ce^{3+}, Ce^{4+}$

$Pt / Sn^{2+}, Sn^{4+}$

(vi) Ion selective electrode (Membrane electrode): In an ion selective electrode, a membrane is in contact with a solution, with which it can exchange ions.

e.g., Glass electrode (exchanges H^+ ions with the solution), various metal ion selective electrodes, which exchange metal ions with the solution.

1.10 REFERENCE ELECTRODES

Reference electrodes are the electrodes with reference to those, the electrode potential of any electrode can be measured. Standard hydrogen electrode is the primary reference electrode as the electrode potential values of other electrodes are assigned with respect to it.

Hydrogen electrode

The hydrogen electrode (Fig. 1.4) consists of a glass tube with a platinised platinum foil fused to it at the bottom. Mercury is placed at the bottom of the tube and a copper wire dipped into it provides

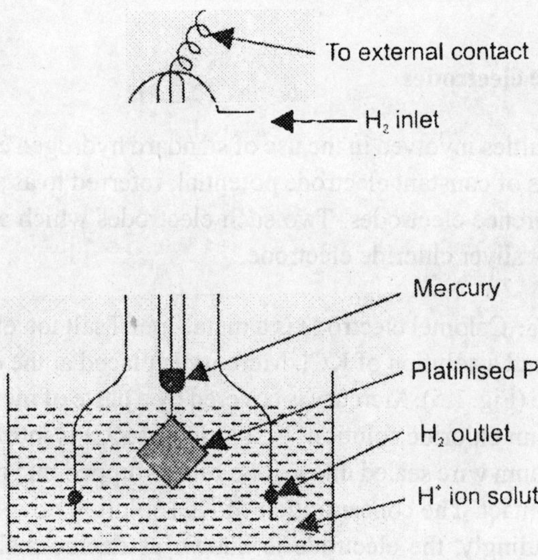

Fig. 1.4. Hydrogen electrode

external electrical contact. The glass tube is surrounded by a bell-shaped glass jacket fused at the top to the glass tube. There is an hydrogen gas inlet at the top. The gas outlets are provided at the bottom.

The platinum foil is dipped in a solution containing H^+ ions and hydrogen gas is bubbled about the electrode through the gas inlet. The hydrogen gas gets adsorbed on the platinum foil and establishes an equilibrium with H^+ ions. The electrode is represented as follows:

$$Pt/H_2(g)/H^+$$

The electrode reaction is: $2H^+ + 2e^- \rightleftharpoons H_2\,(g)$

If the concentration of H^+ ions is 1 M and hydrogen gas is bubbled at 1 atmospheric pressure at 298 K, then it is standard hydrogen electrode. Its electrode potential has been arbitrarily chosen as zero.

Standard hydrogen electrode is used as a primary reference electrode to determine the electrode potential of other electrodes. The test electrode is coupled with a standard hydrogen electrode and the emf of the so formed cell is determined experimentally by potentiometric method. Since the electrode potential of the standard hydrogen electrode is zero, the electrode potential of the test electrode can be calculated.

Limitations

- Construction and working is difficult on account of difficulties involved in maintaining the concentration at unity and in keeping the pressure of the gas uniformly at one atmosphere.
- Platinum is highly susceptible for poisoning by the impurities in the gas.
- It can not be used in the presence of oxidizing agents.

Secondary reference electrodes

Because of the difficulties involved in the use of standard hydrogen electrode as reference electrode, some other electrodes of constant electrode potential, referred to as secondary reference electrodes, are employed as reference electrodes. Two such electrodes which are in common use are calomel electrode and silver – silver chloride electrode.

(i) Calomel electrode: Calomel electrode is a metal-metal salt ion electrode. It consists of mercury, mercurous chloride and a solution of KCl. Mercury is placed at the bottom of a glass tube having a side tube on each side (Fig. 1.5). Mercury is covered by a paste of mercurous chloride (calomel) with mercury and potassium chloride solution. A solution of KCl is introduced above the paste through the side tube. A platinum wire sealed into a glass tube is dipped into mercury and used to provide the external electrical contact. The concentration of KCl solution used is either decinormal, normal or saturated. Correspondingly, the electrode is known as decinormal, normal or saturated calomel electrode, respectively.

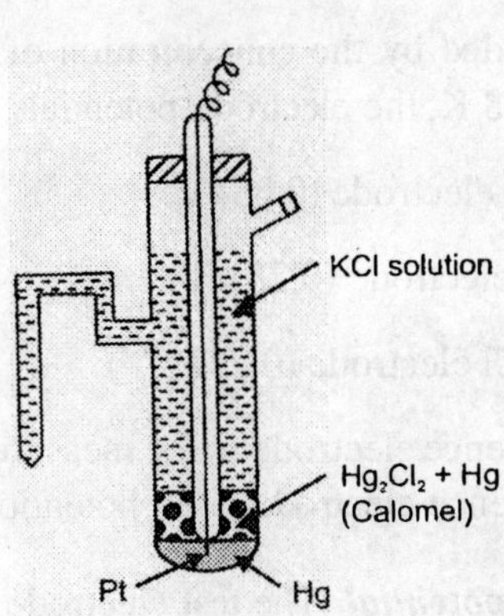

Fig 1.5. Calomel electrode

Calomel electrodes can be represented as

$$Hg/Hg_2Cl_{2(S)}/Cl^-$$

The calomel electrode can act as anode or cathode depending on the nature of the other electrode of the cell.

When it acts as anode, the electrode reaction is,

$$2Hg \rightarrow Hg_2^{2+} + 2e^-$$

$$Hg_2^{2+} + 2Cl^- \rightarrow Hg_2Cl_2$$

$$2Hg + 2Cl^- \rightarrow Hg_2Cl_2 + 2e^-$$

When it acts as cathode, the electrode reaction is,

$$Hg^{2+} + 2e^- \rightarrow 2Hg$$

$$Hg_2Cl_2 \rightarrow Hg_2^{2+} + Cl^-$$

$$Hg_2Cl_2 + 2e^- \rightarrow 2Hg + 2Cl^-$$

The net reversible electrode reaction is,

$$Hg_2Cl_2\,(s) + 2e^- \rightleftharpoons 2Hg\,(l) + 2Cl^-$$

Electrode potential, $E = E^\circ - \frac{2.303\,RT}{2F} \log [Cl^-]^2$

$$E^\circ \quad \frac{2.303\,RT}{F} \log [Cl^-]$$

$$E^\circ - 0.0591 \log [Cl^-] \text{ at } 298\text{ K}$$

The electrode potential is decided by the concentration of chloride ions and the electrode is reversible with chloride ions. At 298 K, the electrode potentials are as follows:

0.1N KCl electrode (0.334V)

1 N KCl electrode (0.281 V)

Saturated KCl electrode (0.2422 V)

Uses: It is used as a secondary reference electrode in the measurement of single electrode potential. It is the most commonly used reference electrode in all potentiometric determinations.

Determination of single electrode potential: The test electrode (e.g., Zn / Zn^{2+}) is coupled with a saturated calomel electrode.

$$Zn(s) / Zn^{2+} // Cl^- / Hg_2Cl_2\ (s) / Hg\ (l)$$

The emf of the so formed cell is determined experimentally by potentiometric method.

$$\begin{aligned} \text{Then } E_{cell} &= E_{cathode} - E_{anode} \\ &= E_{cal} - E_{Zn} \\ &= 0.2422 - E_{Zn} \\ \text{or} \quad E_{Zn} &= 0.2422 - E_{cell} \end{aligned}$$

(ii) Silver-silver chloride electrode: Silver-silver chloride electrode is also a metal-metal salt ion electrode. Silver and its sparingly soluble salt silver chloride are in contact with a solution of chloride ions. Generally a silver wire is coated with AgCl and dipped in a solution of KCl or HCl.

The electrode can be represented as $Ag / AgCl / Cl^-$. The electrode reaction can be represented as follows.

$$AgCl(s) + e^- \rightleftharpoons Ag\ (s) + Cl^-$$

Applying Nernst equation,

$$E = E^\circ - \frac{2.303\ RT}{F} \log [Cl^-]$$ (As [AgCl] and [Ag] are taken as 1]

$$= E^\circ - 0.0591 \log [Cl^-] \text{ at } 298\ K$$

The electrode potential is dependent on the concentration of chloride ions. For 1 molar solution, the electrode potential is 0.223V and for saturated solution it is 0.199V at 298K.

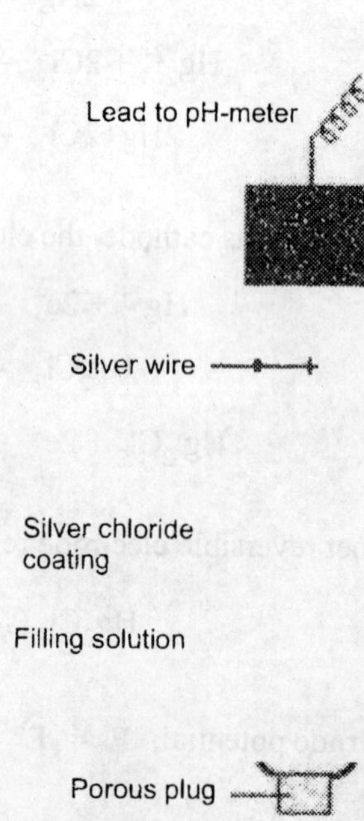

Fig. 1.6. Silver - Silver Chloride Electrode

1.11. ION SELECTIVE ELECTRODES

Ion selective electrode is the one which selectively responds to a specific ion in a mixture and the potential developed at the electrode is a function of the concentration of that ion in the solution. The electrode generally consists of a membrane which is capable of exchanging the specific ions with the solution with which it is in contact. Therefore, these electrodes are also referred to as membrane electrodes.

The various types of ion selective electrodes are listed below.

A. Crystalline membrane electrodes:

1. Single Crystal
   Example: LaF_3 for F^-
2. Polycrystalline or mixed crystal
   Example: Ag_2S for S^{2-} and Ag^+

B. Non-crystalline membrane electrodes:

1. Glass
   Examples: Silicate glasses for Na^+ and H^+
2. Liquid
   Examples: liquid ion exchangers for Ca^{2+}, K^+, etc.
3. Immobilized liquid in a rigid polymer.
   Examples: Polyvinyl chloride matrix for Ca^{2+}, NO_3^-, etc.

An ion selective electrode generally consists of ion selective membrane in contact with an analyte solution on one side and an internal reference solution on the other side. An internal reference electrode is constituted in contact with the reference solution. The electrode can be represented as follows:

Solution to be analysed	Membrane	Internal standard solution	Internal reference electrode
$[M^{n+}] = C_1$	| |	$[M^{n+}] = C_2$	

The mechanism of potential development in ion selective electrodes is different from that in other types of electrodes. The observed potential of an ion selective electrode is a kind of junction potential that develops across a membrane that separates the analyte solution from reference solution. The potential developed across the membrane is a function of concentrations of the analyte and reference solutions.

$$E_j = \frac{2303\,RT}{nF} \log \frac{C_1}{C_2}$$

where C_1 and C_2 are the concentrations of the analyte and reference solution, respectively. If the concentration of the inner standard solution is kept constant, then

$$E_j = k + \frac{2.303\,RT}{nF} \log C_1, \text{ where k is a constant, } k = -\frac{2.303\,RT}{nF} \log C_2$$

The overall potential of the electrode also includes that of the internal reference electrode, and thus,

$$E_m = E_j + E_{ref}$$

$$k + \frac{2.303\,RT}{nF} \log C_1 + E_{ref}$$

$$E^{\circ}_M + \frac{2.303\,RT}{nF} \log C_1, \text{ where } E^{\circ}_M = k + E_{ref}$$

$$\text{At 298K, } E_M = E^{\circ}_M + \frac{0.0591}{n} \log C_1$$

The electrode is combined with an external reference electrode, and emf of the so formed cell is determined by potentiometric method. Knowing the electrode potential of the reference electrode, the electrode potential of the ion selective electrode is calculated, which in turns gives the ion concentrations. The potentiometers are generally automated to read the ion content directly.

Applications

Used to determine the concentrations of a number of cations and anions such as H^+, Li^+, Na^+, K^+, Pb^{2+}, Cu^{2+}, Mg^{2+}, CN^-, NO_3^-, F^-, etc.

The disadvantage with the ion selective electrodes is that, simple potentiometers can not be used with these electrodes, because of the high resistance of the membranes. This necessitates special type of potentiometers which are expensive.

Glass electrode

Glass electrode is a pH sensitive electrode, most widely used for pH determination. A typical glass electrode assembly is shown in Fig. 1.6.

The glass electrode consists of a glass bulb made up of a special type of glass with high electrical conductance. The glass bulb is filled with a solution of constant pH (0.1 M HCl), and is inserted with a Ag – AgCl electrode, which is the internal reference electrode and also serves for the external electrical contact. The electrode is dipped in a solution containing H^+ ions.

Electrode potential of glass electrode

The glass electrode system can be represented as follows:

Ag/AgCl(s) / 0.1M HCl / glass.

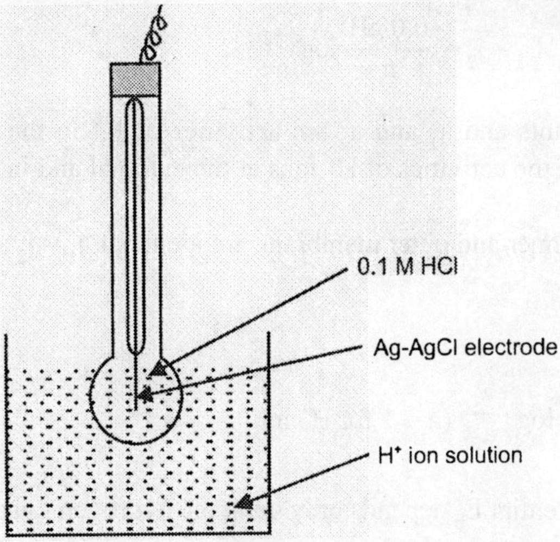

Fig. 1.7. Glass electrode

When dipped in a solution it can be represented as

Glass Membrane

External analyte Solution $[H^+] = C_1$	[ ]	0.1 M HCl $C_2 = 0.1$	Ag/AgCl electrode (Internal ref)
	E_1 E_2		

$$E_b = E_1 - E_2$$

The glass membrane exchanges ions with the solution:

$$H^+_{(Solution)} + MGl_{(Membrane)} \rightleftharpoons M^+_{(Solution)} + HGl_{(Membrane)}$$

The exchange of ions by the inner and outer membrane gives rise to a boundary potential. This boundary potential consists of two potentials E_1 and E_2, which are associated with the outer and inner membranes respectively.

Therefore, $E_b = E_1 - E_2$

From thermodynamic consideration it has been shown that E_1 and E_2 are related to the hydrogen ion activities at each face, by Nernst like equation,

$$E_1 = j_1 - \frac{0.0591}{n} \log \frac{a_1'}{a_1}$$

$$E_2 = j_2 - \frac{0.0591}{n} \log \frac{a_2'}{a_2}$$

where j_1 and j_2 are constants and a_1 and a_2 are activities of H^+ in the outer and inner solution respectively, a_1' and a_2' are the activities of H^+ ions at the external and internal surface of the glass membrane.

Since the glass at the inner and outer membrane are identical, $j_1 = j_2$ and also $a_1' = a_2'$

Then $$E_b = E_1 - E_2$$

$$= 0.0591 \log \frac{a_1}{a_2} \quad (n = 1 \text{ for } H^+ \text{ ion})$$

Thus the boundary potential E_b depends only upon the hydrogen ion activities of the solutions. Since the concentration of the inner solution is constant, a_2 is constant. The above equation simplifies to

$$E_b = L' + 0.0591 \log a_1$$

$$= L' + 0.0591 \log C_1$$

$$= L' - 0.0591 \text{ pH}$$

where $L' = -0.0591 \log a_2 = -0.0591 \log C_2$

The potential of the glass electrode E_G has three components:

(1) the boundary potential

(2) the potential of the internal Ag / AgCl reference electrode and

(3) a small asymmetry potential.

$$E_G = E_b + E_{ref} + E_{asy}$$

$$= L' + 0.0591 \log C_1 + E_{ref} + E_{asy}$$

$$= E°_G + 0.0591 \log C_1$$

$$= E°_G - 0.0591\text{pH}$$

where $E°_G$ is a combination of three constant terms,

$$E°_G = L' + E_{ref} + E_{asy}$$

Asymmetric potential arises due to the difference in responses of the inner and outer surface of the glass bulb to changes in H^+ ion activity. This may originate as a result of differing conditions of strain in the two glass surfaces (due to difference in curvature). This asymmetric potential varies with time and therefore, glass electrode must be standardized frequently by placing a solution of known H^+ activity (pH).

Determination of pH

To determine the pH of a given solution, the glass electrode is dipped in a solution whose pH need to be determined. It is combined with a saturated calomel electrode.

$$Hg / Hg_2Cl_2 / Cl^- // \text{Solution (pH=?)} / \text{glass} / 0.1\ N\ HCl / AgCl / Ag$$

The emf of the so formed cell is determined potentiometrically.

$$E_{cell} = E_G - E_{cal}$$

$$= E°_G - 0.0591\ pH - E_{cal}$$

$$pH = \frac{E°_G - E_{cal} - E_{cell}}{0.0591}$$

$E°_G$ value is evaluated by dipping the glass electrode in a solution of known pH and measuring the emf of the cell formed when combined with a calomel electrode.

Advantages

- Can be used in oxidizing and reducing environments.
- Accurate results are obtained between pH range 1–9. However, by using special glass electrodes, pH 1–14 can be measured.
- Electrode does not get poisoned.

Disadvantages

Because of the high resistance of glass, simple potentiometers cannot be used. It requires sensitive electronic potentiometers for emf measurement.

1.12. CONCENTRATION CELLS

The electrode potential is dependent on the electrolyte concentration, with which the electrode is in contact, as given by Nernst equation. If two electrodes of the same metal are in contact with solutions of its ions with different concentrations, they differ in their electrode potentials. The coupling of two such electrodes also constitutes a cell and gives rise to an emf as the electrodes differ in their electrode potentials. Such cells which consist of two electrodes of the same substance in contact with solutions of different concentrations and the emf arises due to the difference in concentrations are known as concentration cells.

A typical concentration cell is illustrated in Fig. 1.8.

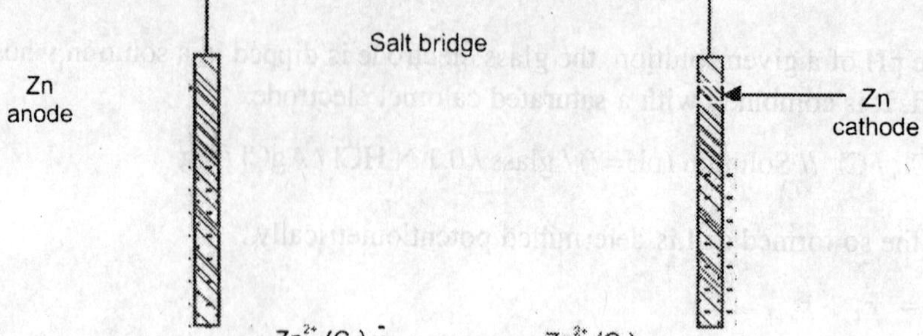

Fig. 1.8. Concentration cell

In this concentration cell, two zinc electrodes arc immersed in zinc sulfate solutions of concentrations C_1 and C_2. The cell is represented as: $Zn / Zn^{2+}(C_1) // Zn^{2+} (C_2) / Zn$.

As in any other cell, oxidation reaction takes place at the anode and reduction takes place at the cathode.

At the anode zinc gets converted into Zn^{2+} ions and goes into the solution, liberating two electrons.

At the anode: $$Zn(s) \rightarrow Zn^{2+} (C_1) + 2e^-$$

At the cathode Zn^{2+} ions are deposited as zinc metal.

At the cathode: $$Zn^{2+} (C_2) + 2e^- \rightarrow Zn (s)$$

There is no net cell reaction. At one electrode zinc atoms go into the solution and at the other equal number of zinc atoms are deposited.

The emf of the ceil is given by,

$$E_{cell} = E_{cathode} - E_{anode}$$

$$= [E^\circ + \frac{2.303\ RT}{nF} \log C_2] - [E^\circ + \frac{2.303\ RT}{nF} \log C_1] \qquad (1.10)$$

$$\frac{2.303\ RT}{nF} \log \frac{C_2}{C_1}$$

It is evident from the above expression that E_{cell} can be positive only if $C_2 > C_1$.

Thus in a concentration cell, the electrode with lower electrolyte concentration acts as the anode and the one with higher electrolyte concentration acts as the cathode. The concentration of ions increases at the anode and decreases at the cathode when the cell is in operation. It is seen from the expression 1.10 that the emf of the cell is dependant on the ratio of electrolyte concentrations at two electrodes. Higher the ratio (C_2 / C_1), higher is the emf. Therefore, the cell can operate only as long as the concentration terms are different. When current is drawn from the cell, C_1 increases and C_2 decreases and as a result the emf of the cell goes on decreasing and becomes zero when the two become equal.

It appears as if the metal ions are transferred from cathode to anode. But there is no direct diffusion of ions from cathode to anode. The metal ions get transferred through electrode reactions.

Concentration cells are also set with two electrodes of the same substance but of different activity in contact with the same solution.

For example,

$$Pt / H_2 (P_{H_2} = P_1) / H^+ / H_2 (P_{H_2} = P_2) / Pt$$

$$Zn (Hg)([Zn]= C_1) / Zn^{2+} / Zn (Hg)([Zn] = C_2)$$

In the first example, hydrogen gases at different partial pressures are bubbled through a same solution of H^+ ions. Since the activities of the gases are different at different pressures, the two electrodes differ in their potentials. In the second example, as the concentrations of zinc in two amalgams are different, their activities are also different though they are in contact with the same solution. This gives rise to difference in their electrode potentials.

Example 1.7. A concentration cell is constructed by dipping copper rods in 0.001M and 0.1M $CuSO_4$ solutions. Calculate the emf of the cell at 298 K.

Solution:

The electrode with lower electrolyte concentration acts as anode and the other acts as cathode.

The cell is represented as, $Cu(s) / Cu^{2+} (0.001) // Cu^{2+} (0.1) / Cu(s)$

At the anode, $Cu \rightarrow Cu^{2+} + 2e^-$

At the cathode, $Cu^{2+} + 2e^- \rightarrow Cu$

$$\frac{0.0591}{n} \log \frac{C_2}{C_1} \quad (C_2 > C_1) \text{ at } 298 \text{ K}$$

$$\frac{0.0591}{2} \log \frac{0.1}{0.001}$$

$$\frac{0.0591}{2} \log 100$$

0.0591 V

Example 1.8. Calculate the emf of the following concentration cell at 298K.

$$Ag(s) / Ag^+(0.01) // Ag^+(0.5) / Ag(s)$$

Solution:

At the anode. $Ag \rightarrow Ag^+ + e^-$

At the cathode, $Ag^+ + e^- \rightarrow Ag$

$$\frac{0.0591}{n} \log \frac{C_2}{C_1} \quad (C_2 > C_1) \text{ at } 298 \text{ K}$$

$$\frac{0.0591}{1} \log \frac{0.5}{0.01}$$

0.101 V

REVIEW QUESTIONS

1. Define the following:
   (i) Electrochemical cell
   (ii) Galvanic cell
   (iii) Electrolytic cell
2. Explain the following terms with respect to a galvanic cell:
   (i) Anode
   (ii) Cathode
   (iii) Half cell
   (iv) Half cell reactions (iv) Cell reaction.
3. Define emf of a cell. How is it determined?
4. Define the following:
   (i) Single electrode potential
   (ii) Standard electrode potential.
5. Explain the origin of electrode potential.
6. What are the sign conventions for electrode potential?
7. Derive Nernst equation for electrode potential.
8. Explain the function of a salt bridge.
9. With examples mention different types of electrodes.
10. What is a primary reference electrode? What are its limitations?
11. Explain the construction and working of the following:
    (i) Calomel electrode
    (ii) Ag – AgCl electrode.
12. Explain a method for the determination of single electrode potential.
13. How do the electrode potential values explain the reactivities of different metals?
14. What are ion selective electrodes? Why are they also known as membrane electrodes?
15. Explain the working of a ion selective electrode.
16. Explain the construction and working of a glass electrode. Mention their advantages and disadvantages.
17. Derive an expression for the electrode potential of a glass electrode.
18. How can glass electrode be used for pH determination of a solution?

19. Define a concentration cell. Explain the working of a concentration cell.
20. Justify the following statements:
    (i) Voltmeter can not be used to measure the emf of a cell.
    (ii) In a galvanic cell anode is –ve and cathode is +ve.
    (iii) Electrode potential values are said to be in hydrogen scale.
    (iv) EMF of a cell is always positive.
    (v) Calomel electrode is reversible with chloride ions.
    (vi) Zinc displaces copper from $CuSO_4$ solution but silver does not.
    (vii) Magnesium displaces hydrogen from acid but mercury does not.
    (viii) Reactivity of a metal inversely varies with its electrode potential.
    (ix) From a concentration cell, emf is available only as long as the electrolytes differ in their concentrations.

CHAPTER 2

Battery Technology

2.1. INTRODUCTION

The developments in the field of science, engineering and technology naturally tempt man in using and applying these developments to improve the comforts in life and enjoyment. The increase in complexity and standards in life style is characterized by a high demand for energy consumption. Even if energy is available, it should be available at a time, place, and under circumstances where desired or needed. It is, in this view point, cells and batteries are important in providing portable electrical energy. Transistorized and miniaturized equipments have such low power requirements that batteries have become competitive with other energy sources, in spite of their lower energy production. The variety of equipments used in diverse applications, by utilizing batteries as energy sources have spurred the development of many different types of batteries.

About six decades ago domestic uses of batteries were largely confined to flash lamps, radio sets and starter batteries for cars and motor cycles. Modern households typically have 40–50, hidden away in all sort of consumer products – from clocks and watches to personal CD players and mobile phones. Away from the home there are many other applications, particularly for large batteries. Examples include the standby batteries for emergency use in hospitals, hotels, departmental stores, telephone exchanges, etc; traction batteries for electric vehicles (tugs, tractors, wheel chairs, golf carts); batteries for solar panels or wind generators; defence batteries in armaments, missiles, submarines, torpedoes. Many of these applications demand a performance that is barely matched by traditional batteries and therefore, there is ever present demand for new and better varieties. One of the interesting features of batteries is the very wide range of sizes in which they are manufactured, from a stored energy content of 0.1 watt-hour (Wh) for a watch or calculator battery, to 100 MWh for a load-levelling battery in the electrical supply industry.

The conversion of chemical energy into electrical energy is the basis for the functioning of a galvanic cell, named after its inventor Luigi Galvani (1786). "Voltaic piles" of unlike metals in contact with an electrolyte, developed by Alessandro Volta (1800) was the first power source. A major advance in the evolution of the battery was the Daniell cell invented by J.F. Daniell (1836). Plante (1859) developed the essentials of the lead-sulfuric acid storage cell. In 1868 George Leclanche introduced a cell which was the forerunner of the present day dry cell. Because of the chemical similarities, the dry cell, is still referred to as a Leclanche-type cell. The first true dry cell was developed between 1886 and 1888 by Dr. Carl Gassner. T.A. Edison, Adams, Ruben, are the few others who have contributed to the development of battery technology.

The battery technology, historically considered electrochemical, combines the activities of the chemist, the chemical engineer, the metallurgist, the materials specialist, the electrical engineer, the electronic designer, the space specialist, the environmental engineer, the instrumental analyst, testing agencies as well as the application originator for expanded and new uses. The individual viewpoints of all these, cooperate to serve the diversified needs and desires of industries, as well as our daily lives in the electronic and space age.

2.2. BATTERIES AND THEIR IMPORTANCE

The conversion of chemical energy into electrical energy is a function of cells or batteries. A cell designates a single unit. A battery is an arrangement of two or more cells, usually connected in series or parallel to supply the necessary current or voltage or both.

The battery was the first practical source of electrical energy developed in man's search for portable power sources. Although many other techniques have been developed for supplying electrical power, the battery is still the most widely used source of electrical power when portability is the prime requisite. Batteries are used in watches, hearing aids, flash lights, radio sets, electric clocks, photoflash devices, emergency lighting, distress signalling, rescue devices, tape recorders, alarm systems, hand tools, toys, calculators, cordless appliances, telephone systems, electrotherapeutic purposes, electric bells, motor ignitions, space vehicles, military applications and a host of other applications. The development of semi conductor devices such as transistors, diodes, missiles, satellites and a great variety of mobile equipments, has imposed great demands for power sources which are compact, dimensionally adaptable, able to operate over a wide temperature range and highly dependable. The batteries meet these demands and therefore, are in continuous demand.

2.3. COMMERCIAL CELLS

The spontaneous redox reaction, which forms the basis of voltaic cell, is conveniently used in making many useful commercial cells. All voltaic cells do not find commercial applications. A useful commercial cell should meet the following basic requirements or specifications:

- Portability
- Should be compact and light weight
- Should provide economically priced, continuous electric supply
- High cell voltage and stable voltage plateau over most of the discharge
- High stored energy content per unit mass and per unit volume
- Low cell resistance
- High sustained power output
- Wide temperature range of operation
- Long inactive shelf life and long operational life.

- Low initial cost
- Sealed and leak proof
- Rugged and resistant to abuse
- Safe in use and under accident conditions
- Should be capable of recharging with a capacity of many charge and discharge cycles, and to withstand overcharge and over discharge
- Sealed and maintenance free

In spite of their virtues, the high cost of cells and batteries makes the production of electricity in large quantities impracticable. However, they serve as convenient energy sources where small amount of energy is required for short duration of time.

Classification

The galvanic or voltaic cells used as source of electrical energy are of two types:

(i) primary cells and
(ii) secondary cells.

(i) Primary cells: A primary cell is the one in which electrical energy can be obtained at the expense of chemical energy only as long as the active materials are still present. Once these have been consumed, the cell can not be profitably or readily rejuvenated and must be discarded. Or in other words, they can not be recharged and re-used.

(ii) Secondary cells: A secondary cell, once used can be recharged by passing current through it. It can be used over and over again. The redox reaction gets reversed during recharging. Since the electrical energy brings about the chemical change, it is converted into chemical energy. Thus electrical energy is stored in the form of chemical energy and utilized for supplying the current when needed. Secondary cells are also known as storage cells.

Whereas, a primary cell acts only as a galvanic or voltaic cell, a secondary cell can act both as galvanic cell and electrolytic cell. During discharging it acts as a galvanic cell, converting chemical energy into electrical energy and during charging it acts as an electrolytic cell converting electrical energy into chemical energy.

Reserve Batteries

In this type of batteries, a key component is separated from the rest of the battery prior to activation. Usually the electrolyte is the component that is isolated. Batteries which use highly active component material are designed in this form to withstand deterioration in storage and to eliminate self discharge prior to use. The reserve design is also used for batteries required to meet extremely long or environmentally severe storage requirements.

When one of the key components of the cell is separated from the remainder of the cell, chemical reaction between the cell components (self discharge) is prevented and the battery is capable of long term storage. The electrolyte is the component that is usually isolated. But in some water activated batteries, the electrolyte solute is contained in the cell and only water is added for activation. Examples of reserve batteries are magnesium – water activated batteries, zinc-silver oxide reserve batteries, etc. Thermal batteries are reserve batteries that use inorganic salt electrolytes, which are non-conductive solids. The battery is activated by melting the electrolyte.

Cell Characteristics

The suitability of a battery system for a particular application is decided based on certain performance characteristics of the battery or the cells constituting the battery system. Some of the important characteristics are briefly discussed below:

(i) Voltage: The voltage available from a battery depends upon the emf of the cells which constitute the battery system. The emf of the cell depends on the free energy change in the overall cell reaction. As given by Nernst equation,

$$E_{cell} = E^{o}_{cell} - \frac{2.303\ RT}{nF} \log Q,$$

where $E_{cell} = E_{cathode} - E_{anode}$, and Q is the reaction quotient for the cell reaction at any stage of the reaction, which is the ratio of the product of molar concentration of the reaction product molecules to that of the reactants.

As it is evident from the above equation, emf of the cell and also the voltage available from the battery is dependent on standard electrode potential difference between the cathode and the anode, temperature and the extent of the cell reaction.

- If the difference in the standard electrode potential is more, higher is the emf of the cell.
- As the temperature increases the emf of the cell decreases.
- As the value of Q increases, i.e., when current is drawn from the cell and cell reaction proceeds, the emf of the cell decreases marginally.
- Also, high cell potential is possible when the cell is with low resistance, i.e., with a high conductivity electrolyte.

(ii) Current: Current is a measure of the rate at which the battery is discharging. High current can be delivered without excessive voltage penalty if there is rapid electron transfer reaction.

(iii) Capacity: The capacity is the charge or the amount of electricity that may be obtained from the battery and is given in ampere hours (Ah). Capacity depends on the size of the battery, and is given by Faraday's relation

$$C = \frac{wnF}{M}$$

where, w is the mass and M is the molar mass of active material. The ability of the cell to continue to deliver the expected capacity at increased discharge rates is an important battery property.

(iv) Electricity storage density: Electrical storage density is the amount of electricity per unit weight which the storer can hold, i.e., it is the capacity per unit weight of the battery.

(v) Power density: Power density is the power per unit weight of battery and is given by iE_{cell} per unit weight of battery. The power density decreases during discharge.

(vi) Cycle life: Primary batteries are designed for single discharge, but a secondary battery is rechargeable. The cycle life is the number of charge / discharge cycles that are possible before failure occurs. The cycle life of a storage battery must be high.

(vii) Energy efficiency: The energy efficiency of a secondary battery is given by

$$\%\text{Energy efficiency} = \frac{\text{energy released on discharge}}{\text{energy required for charge}} \times 100$$

Higher the energy efficiency, better is the battery.

(viii) Shelf life: It is essential for most batteries to be stored, sometimes for many years, without self discharge or corrosion of current collectors, causing a loss of performance.

(ix) Tolerance to service conditions: The battery has to be tolerant to different service conditions such as variation in temperature, vibration and shock.

2.4. PRIMARY CELLS

Dry Cell

The dry cell (Fig. 2.1) consists of a cylindrical zinc container, acting as the anode and a graphite rod placed at the centre of the cylinder acting as the cathode. The anode is in contact with a paste of ammonium chloride and zinc chloride as electrolyte. Cathode carbon rod is surrounded with a paste of manganese dioxide (MnO_2) and carbon powder. As the conductivity of MnO_2 is low, granulated carbon, more or less completely graphitized, is added to increase conductivity.

Electrode reactions

The primary electrode reactions are:

At the anode, $Zn \rightarrow Zn^{2+} + 2e^-$

At the cathode, $2MnO_2 + H_2O + 2e^- \rightarrow Mn_2O_3 + 2OH^-$

The overall cell reaction is, $Zn + 2MnO_2 + H_2O \rightarrow Zn^{2+} + Mn_2O_3 + 2OH^-$

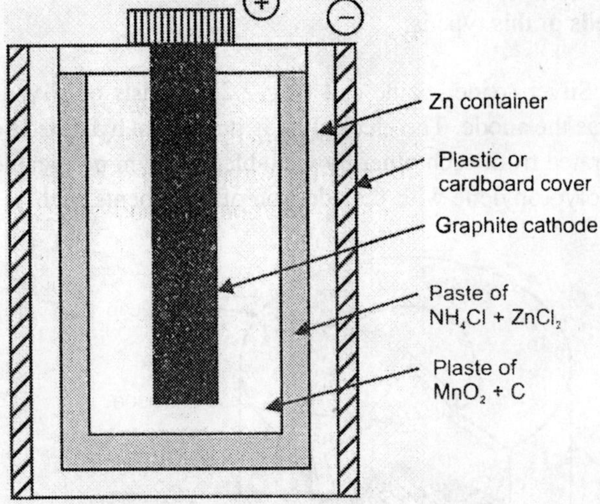

Fig. 2.1. Dry cell

Secondary reactions

The hydroxyl ions generated during the cell reaction liberate ammonia from ammonium chloride, which in turn combines with the Zn^{2+} ions to precipitate $[Zn[NH_3]_2Cl_2]$ complex.

$$2NH_4Cl + 2OH^- \rightarrow 2NH_3 + 2H_2O + 2Cl^-$$

$$Zn^{2+} + 2NH_3 + 2Cl^- \rightarrow [Zn\,[NH_3]_2\,Cl_2]$$

The secondary reactions are not involved directly in the electrode reaction and so they do not contribute to the emf of the cell. But these reactions are irreversible and therefore, the cell can not be recharged.

A fresh dry cell has a potential of l.5V. The voltage of the cell decreases gradually with usage and finally it has to be discarded.

Applications

Dry cells find extensive use in flash lights, portable radios, transistors, tape recorders, and similar electronic devices where small amount of current is required.

Limitations

- They have low capacity
- Not intended for heavy duty

(ii) Alkaline primary cells: In alkaline primary cells the electrolyte is an alkali like NaOH or KOH. One of the important cells of this type is:

Silver oxide-Zinc cell: Silver oxide – zinc cell (Fig 2.2) consists of silver oxide (Ag_2O) as the cathode and zinc metal as the anode. The electrolyte is potassium hydroxide or sodium hydroxide. The electrodes are separated from each other by wettable and semi permeable organic membranes such as cellophane or polypropylene with the addition of absorbents such as rayon or nylon.

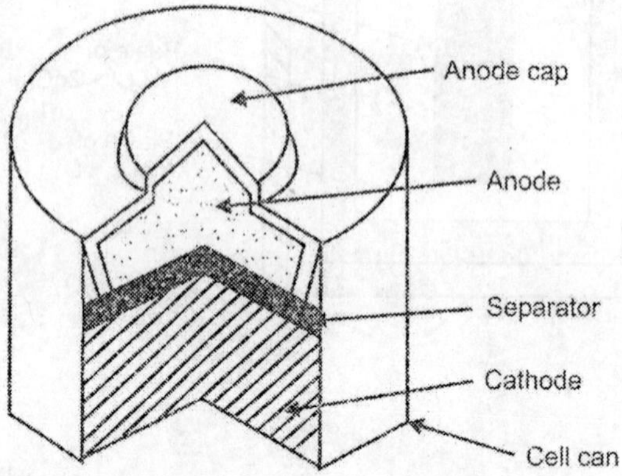

Fig. 2.2. Silver oxide – Zinc cell

Electrode reactions:

At the anode: $Zn + 2OH^- \rightarrow ZnO + H_2O + 2e^-$

At the cathode: $Ag_2O + H_2O + 2e^- \rightarrow 2Ag + 2OH^-$

Overall cell reaction is, $Zn + Ag_2O \rightarrow ZnO + 2Ag$

In this cell also, the electrolyte KOH is non-variant.

The cell gives an emf of 1.5V. The cell has a high energy output per unit weight and a fairly constant voltage level during discharge.

Applications:

- In military applications.
- It is a miniature dry cell used in hearing aids, electronic watches, photoelectric exposure devices and reference voltage sources.

2.5. SECONDARY CELLS OR STORAGE CELLS (ACCUMULATORS)

Secondary cells or storage cells are rechargeable cells on exhaustion of reactive chemicals in the cell. They can be classified into two types:

(i) Acid storage cells and

(ii) Alkaline storage cells.

While lead storage cell is an acid storage cell, nickel cadmium cell is an alkaline storage cell.

(i) Acid storage battery (Lead storage battery)

A simplified form of lead storage cell is as shown in Fig 2.3. The electrodes are lead grids. The anode grid is filled with spongy lead and the cathode grid is filled with lead dioxide (PbO_2). A number of electrode pairs with inert porous partitions in between, are dipped in approximately 20% sulfuric acid, which is the electrolyte. The battery is encased in a plastic container.

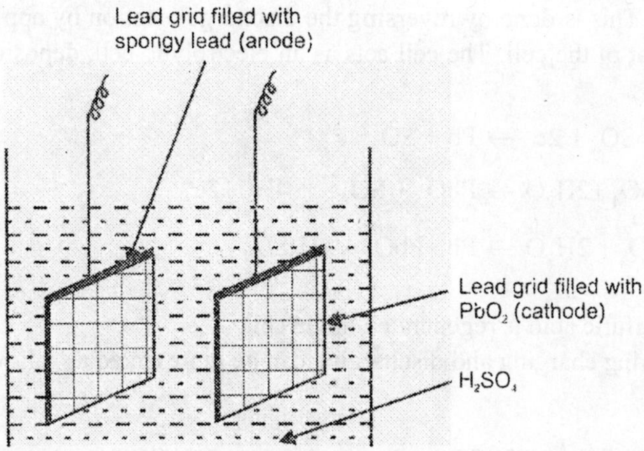

Fig. 2.3. Lead Storage Battery

Electrode reactions: The electrode reactions that occur during the discharge of the cell, i.e., on drawing current from the cell, are as follows:

At the anode,

$$Pb \rightarrow Pb^{2+} + 2e^-$$

$$Pb^{2+} + SO_4^{2-} \rightarrow PbSO_4$$

$$Pb + SO_4^{2-} \rightarrow PbSO_4 + 2e^-$$

At the Cathode,

$$PbO_2 + 4H^+ + 2e^- \rightarrow Pb^{2+} + 2H_2O$$

$$Pb^{2+} + SO_4^{2-} \rightarrow PbSO_4$$

$$PbO_2 + 4H^+ + SO_4^{2-} + 2e^- \rightarrow PbSO_4 + 2H_2O$$

The net cell reaction is,

$$Pb + PbO_2 + 2H_2SO_4 \rightarrow 2PbSO_4 + 2H_2O$$

The product ($PbSO_4$) formed during discharge remains adhered on each electrode and is available at the site during recharging.

From the cell reaction, it is evident that the potential of the lead storage cell is dependent at any given temperature on the activity or concentration of sulfuric acid solution. Thus at 25°C and at a concentration of 7.4% sulfuric acid the potential developed by a pair of electrode is 1.90V; at 21.4%, 2.0V and at 39.2% 2.14V. Further, it is evident that sulfuric acid is consumed and water is produced during discharge. This results in the decrease in the specific gravity of the electrolyte.

Recharge reactions: When the density of sulfuric acid in the battery falls below 1.20gcm^{-3}, the battery needs charging. This is done by reversing the discharge reaction by applying externally a potential higher than that of the cell. The cell acts as an electrolytic cell, depositing lead and lead dioxide on the electrodes.

$$PbSO_4 + 2e^- \rightarrow Pb + SO_4^{2-}$$

$$PbSO_4 + 2H_2O \rightarrow PbO_2 + SO_4^{2-} + 4H^+ + 2e^-$$

$$2PbSO_4 + 2H_2O \rightarrow Pb + PbO_2 + 2H_2SO_4$$

During charging, sulfuric acid is regenerated in the cell.
The net reaction during charging and discharging can be represented as follows:

$$Pb(s) + PbO_2(s) + 2H_2SO_4(aq) \underset{\text{Charging}}{\overset{\text{Discharging}}{\rightleftharpoons}} PbSO_4(s) + 2H_2O\,(l)$$

Applications: Each electrode pair develops a potential of 2V. In order to obtain higher potential a number of electrode pairs are connected in series.

Lead storage batteries are extensively used in automobiles to start the engine. They are also used for electric supply in telephone exchangers, railway trains, hospitals, laboratories, etc., and for emergency power supplies.

Limitations:

- The potential decreases with decrease in concentration of sulfuric acid.
- Excessive discharge and quick charging shortens the life of the battery.
- Cell potential and the effectiveness is reduced at low temperature.
- Excessive charging may damage the electrodes and may also lead to explosion.

(ii) Alkaline storage battery: Nickel-cadmium cell

Nickel cadmium cell consists of a nickel wire gauze electrode grids. The anode grid consists of a mixture of spongy cadmium with 78% cadmium hydroxide, 18% iron, 1% nickel and 1% graphite. The cathode grid contains nickel hydroxide (80%), cobalt hydroxide (2%), graphite (18%) and traces of barium compound. Graphite increases the conductivity, the cobalt and barium compounds increases the efficiency of active material and also the cycle life. 6M KOH is the electrolyte.

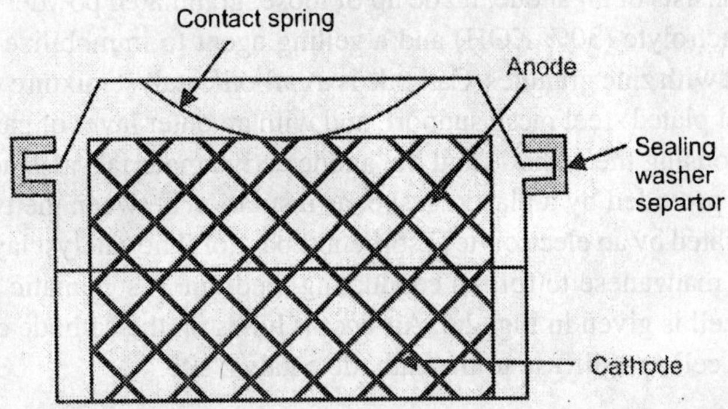

Fig. 2.4. Button type Ni-Cd cell

In a button type cell (Fig. 2.4), a cell cup acts as the cathode and a cell cover acts as anode. The electrodes consist of tablets wrapped in nickel wire gauze, separated by a fine porous separator soaked with the electrolyte. Sealing is accomplished by flanging the rim of the cell cup over the rim of the cell cover with a plastic washer in between, serving to insulate the cup from the cover.

Electrode reactions: During discharge,

At the anode, $Cd + 2OH^- \rightarrow Cd(OH)_2 + 2e^-$

At the cathode, $2Ni(OH)_3 + 2e^- \rightarrow 2Ni(OH)_2 + 2OH^-$

The net cell reaction is

$$Cd + 2Ni(OH)_3 \rightarrow 2Ni(OH)_2 + Cd(OH)_2$$

During charging, the above reaction is reversed.

The open circuit potential of the cell ranges from 1.4 to 1.28V. It is compact, light weight battery with good cycle characteristics, capacity and long shelf life.

Applications: Nickel cadmium cells are used in battery operated appliances such as pocket calculators, photo flash units, cordless garden tools, electric shavers, instruments, alarm systems, transmitters, receivers, emergency lighting, hearing aids, telemeters, etc.

2.6 MODERN BATTERIES

(i) Zinc-air cell

Zinc-air battery is a type of metal – air batteries, which use oxygen directly from the atmosphere to produce electrochemical energy. Oxygen diffuses into the cell and is used as the cathode reactant. The air cathode catalytically promotes the reaction of oxygen with an alkaline electrolyte and is not consumed or changed during discharge. As the cathode can be very compact, high energy densities are achieved.

The zinc-air cell consists of an anode, made up of loose, granulated powder of zinc mixed with an aqueous alkaline electrolyte (30% KOH) and a gelling agent to immobilize the composite and ensure adequate contact with zinc granules. Cathode is a carbon/catalyst mixture with a wet proofing agent coated on a nickel plated steel mesh support, and with an outer layer of gas permeable Teflon layer. The can halves housing the cathode and the anode active materials also act as the terminals. The two containers are provided by a plastic gasket as insulation between the two containers. The two electrodes are separated by an electrolyte absorbent separator. The catalyst layer contains carbon blended with oxides of manganese to form a conducting medium. A schematic representation of a typical zinc air button cell is given in Fig. 2.5. Air access holes on the cathode can provides a path for oxygen to enter the cell and diffuse to the cathode catalyst site.

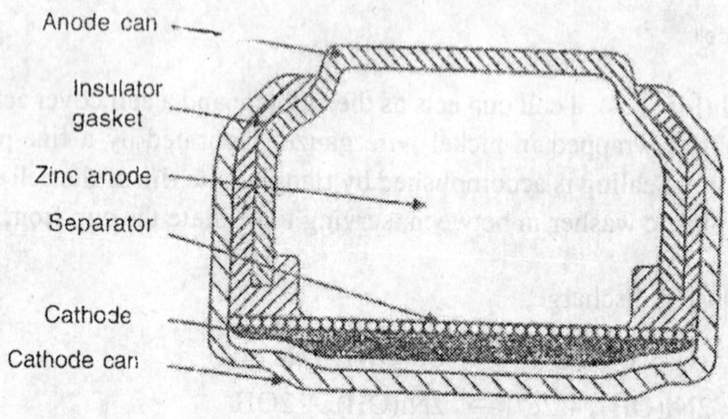

Fig. 2.5. Zinc-air cell

Electrode reactions:

At the anode,

$$Zn \rightarrow Zn^{2+} + 2e^-$$

$$Zn^{2+} + 2OH^- \rightarrow ZnO + H_2O$$

$$Zn + 2OH^- \rightarrow ZnO + H_2O + 2e^-$$

At the cathode,

$$\frac{1}{2} O_2 + H_2O + 2e^- \rightarrow 2OH^-$$

Cell reaction:

$$Zn + \frac{1}{2} O_2 \rightarrow ZnO$$

The cell produces an open circuit potential of 1.4V.

During the cell reaction the electrolyte remains invariant and the air cathode acts only as a reaction site and is not consumed. A very thin cathode of the cell (about 0.5 mm) permits the use of very large zinc anode. This results in higher energy density.

Advantages

- High energy density
- Long shelf life (sealed)
- No ecological problems
- Low cost

Disadvantages

- Limited power out put
- Short activated life

Applications

- Used as a power source for hearing aids.
- Used in electronic pagers.
- Used in telemetry/voice transmitters.
- Used in various medical devices.
- Large zinc – air batteries used in railroad signaling, seismic telemetry, remote communications, etc.

(ii) Nickel-metal hydride batteries

The rechargeable nickel-metal hydride battery is similar in its characteristics to nickel-cadmium battery. The main difference is that nickel metal hydride battery uses hydrogen, adsorbed in a metal alloy for the active negative material in place of the cadmium in the nickel-cadmium battery.

The active material at the cathode of the nickel- metal hydride battery is nickel oxyhydroxide (NiOOH) and at the anode is hydrogen in the form of metal hydride. The metal alloy is capable of undergoing a reversible reaction as the battery is charged or discharged. Two types of metal alloys are used – AB_5 class of alloys consisting of lanthanum and nickel, and AB_2 class of alloy, consisting of titanium and zirconium. In both cases some of the base metals are replaced with other metals to achieve the desired characteristics. An aqueous solution of potassium hydroxide is used as the electrolyte.

The cathode in the nickel-metal hydride cell is a highly porous sintered or felt nickel substrate into which the nickel oxyhydroxide is impregnated or pasted. The anode is a highly porous structure with a nickel wire gauze grid into which the plastic bonded active hydrogen storage alloy is coated. The electrodes are separated with a synthetic non woven material, which serves as an insulator between the two electrodes and as a medium for absorbing the electrolyte. A button type cell is shown in Fig. 2.6.

Electrode reactions: During discharge,

At the anode, $MH + OH^- \rightarrow M + H_2O + e^-$

At the cathode, $NiOOH + H_2O + e^- \rightarrow Ni(OH)_2 + OH^-$

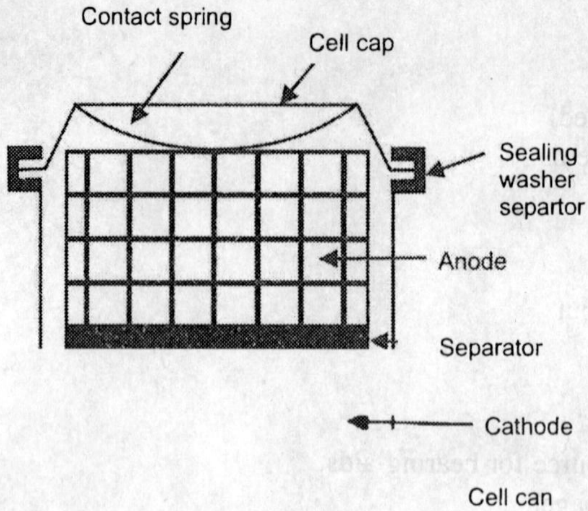

Fig. 2.6. Nickel-metal hydride button cell

The net cell reaction is, $MH + NiOOH \rightarrow M + Ni(OH)_2$

During charging, the above reaction is reversed.
The open circuit potential of the cell ranges from 1.25 – 1.35 V.

Advantages:

- High capacity.
- Sealed construction, no maintenance required.
- Cadmium free – minimal environmental problems.
- Rapid recharge capability.
- Long cycle life.
- Long shelf life.

Disadvantages: Performance is not as good as with nickel-cadmium batteries.

Applications:

- In consumer electronic devices such as cellular phones, computers, comcorders and other portable applications.
- In electric vehicles.

(iii) Lithium batteries

Lithium metal offers an attractive option be used as a battery anode material because of its light weight, low electrode potential, high electrochemical equivalence and good conductivity. For these reasons, use of lithium has predominated in the development of high performance, high energy density primary and secondary batteries.

The term lithium battery refers to a large family of batteries with a common feature of having lithium negative electrode (anode). The different lithium battery systems differ in the choice of electrolyte medium and of the positive electrode system.

Classification: Lithium batteries are classified into primary and secondary batteries. Primary battery is not chargeable and the secondary battery is chargeable.

Based on the type of cathode material used, lithium primary cells are classified as follows:

(a) Soluble-cathode cells: In this type of cells liquid or gaseous cathode materials, such as sulphur dioxide or thionyl chloride, are used. These substances dissolve in the electrolyte or are the electrolyte solvent.

(b) Solid-cathode cells: These types of cells use solid material for the cathode substances, such as V_2O_5, MnO_2, CuS, Fe_2S_3, CuO, etc.

(c) Solid electrolyte cell: This type of cells use electrolytes in the solid form it self as the cathode. PbI_2, PbS, etc., are used as solid electrolyte cathodes.

The electrolytes used in lithium batteries can not be aqueous solutions, because of the high reactivity of lithium with water. Therefore non aqueous electrolytes are to be used in lithium batteries. The different types of electrolytes used are of the following types:

- Lithium salt solutions in organic solvents such as propylene carbonate, dioxolane, THF, ethers, acetonitrile, etc.
- Solvents like thionyl chloride, sulfuryl chloride, which are also the electro active species at the cathode, mixed with lithium salts to provide ionic conductivity.
- A lithium ion conducting solid electrolyte such as LiI or LiI + Al_2O_3
- An organic polymer, which can conduct lithium ions. e.g., Polyethylene oxide.
- A molten salt. e.g., Molten LiCl + KCl

Lithium batteries based on the combinations of the above types of solvent-electrolyte system and positive electrodes have been produced and tested.

Advantages of Lithium batteries

- High cell voltage, up to above 4 V, depending on the cathode material. This is because of the very negative electrode potential of Li/ Li^+.
- High energy density due to the low atomic mass of lithium. 1F is released by the dissolution of 7 g of the metal.
- Operation over a wide temperature range, from about 70 to – 40° C.
- Flat discharge characteristics-constant voltage and resistance through most of the discharges of many lithium cells.
- Superior shelf – life.

Lithium-Manganese dioxide cell: Lithium-Manganese dioxide (Li / MnO_2) cell is a solid cathode lithium cell. It consists of a lithium anode and manganese dioxide cathode. A solution of lithium halide in organic solvent serves as the electrolyte. A coin cell is shown in Fig. 2.7.

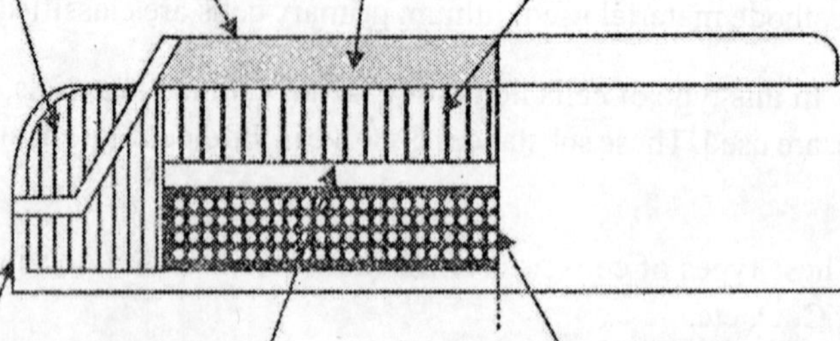

Fig. 2.7. Lithium-MnO_2 button cell

The manganese dioxide pellet and lithium anode disc are separated by a non woven polypropylene separator impregnated with the electrolyte. The cell is sealed with the can serving as the cathode terminal and the cap as the anode terminal.

Electrode reactions:

At the anode, $Li \rightarrow Li^+ + e^-$

At the cathode, $MnO_2 + Li^+ + e^- \rightarrow LiMnO_2$

The net cell reaction is, $Li + MnO_2 \rightarrow LiMnO_2$

Mn^{4+} is reduced to Mn^{3+}, with the Li^+ entering the MnO_2 crystal lattice.

Applications

- As a long time memory backup
- In watches, calculators, cameras, lighting equipments, toys and other consumer electronics.

Rechargeable lithium batteries: Rechargeable lithium batteries offer advantages like high energy density, higher cell voltage and longer charge retention or shelf life, when compared to conventional aqueous secondary cells. However, they have disadvantages such as low cycle life, relatively poor low temperature performance and capacity fading. A variety of cathodic materials and electrolytes are being used in these secondary cells as in the case of primary lithium batteries. A classification with components, characteristics and examples is given in Table 2.1.

Table 2.1. Classification of rechargeable lithium batteries

Cell	*Components*	*Characteristics*	*Examples*
1. Liquid organic electrolyte cells	**Anode:** Lithium **Cathode:** Solid cathode using intercalation compounds. **Electrolyte:** Liquid organic electrolyte	High specific energy Moderate rate capacity Low cycle life Low self discharge rate	Li / MoS_2 Li / MnO_2 Li / TiS_2 Li / V_2O_5 Li / $LiNiO_2$
2. Polymer electrolyte cells	**Anode:** Lithium **Cathode:** Intercalation compounds **Electrolyte:** Polymer electrolyte	High energy density Safer design Low electrolyte conductivity Poor low temperature performance Low self discharge rate	Li / PEO $LiClO_4$ / V_6O_{13}
3.Lithium ion cells	**Anode & Cathode:** Intercalation Compounds **Electrolyte:** Liquid organic or polymer electrolyte	Safest cell design Good cycle life Poor charge retention High self discharge rate	Li_xC / $LiCoO_2$ Li_xC / $LiNiO_2$ Li_xC / $LiMnO_2$
4. Inorganic electrolyte cells	**Anode:** Lithium **Cathode:** Liquid cathode of inorganic materials **Electrolyte:** Cathodic material itself	High energy density Excellent shelf life Ability to sustain overcharge Safer	Li / SO_2 Li / $CuCl_2$
5. Lithium alloy cells	**Anode:** Lithium alloy **Cathode:** Variety of types **Electrolyte:** Liquid organic electrolyte	Safer Low energy density Poor cycle life Coin cell configuration	LiAl / MnO_2 LiAl / V_2O_5 LiAl / C LiAl / Polymer

Lithium ion batteries: A lithium-ion battery is a rechargeable battery type in which lithium ions move from the negative electrode to the positive electrode during discharge and back when charging. Li-ion batteries use an intercalated lithium compound as the electrode material, compared to the metallic lithium used in non-rechargeable lithium battery.

The anode is made of lithium, intercalated (reversibly included between the layers) in graphite. The cathode is made up from lithium liberating compounds, typically the three electro-active layered oxide materials,

- Lithium cobalt-oxide ($LiCoO_2$)
- Lithium manganese-oxide ($LiMn_2O_4$)
- Lithium nickel-oxide ($LiNiO_2$).

The **electrolyte** is a solution of lithium salt (such as lithium hexafluorophosphate, lithium perchlorate, lithium tetrafluoroborate, etc.) in an organic solvent such as ethylene carbonate, dimethyl carbonate, and diethyl carbonate. The electrodes used in these cells have two key properties. One is the open crystal structure, which allows the insertion or extraction of lithium ions and the second is the ability to accept compensating electrons at the same time. Such electrodes are called intercalation hosts. The construction of a lithium ion cell is shown in Fig. 2.8.

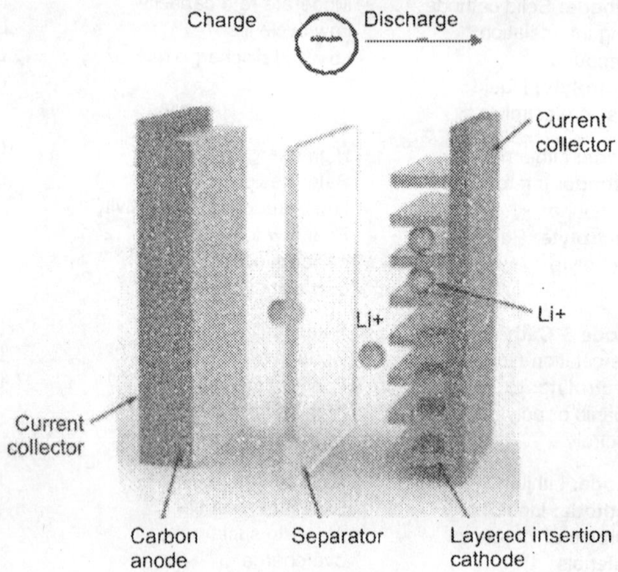

Fig. 2.8. Lithium ion cell

Li-ion cell has a four-layer structure. A positive electrode, **cathode** is made with lithium cobalt oxide, has a current collector made of thin aluminium foil. A negative electrode, **anode** is made with graphite, has a current collector of thin copper foil. A **separator** is a fine porous polymer film. An **electrolyte** is made with lithium salt in an organic solvent. The electrolyte provides a conductive medium for lithium ions to move between the electrodes.

Both the electrodes allow lithium ions to move in and out of their interiors. During charging, lithium in cathode electrode material is ionized and moves from cathode to anode and gets inserted into the anode. During discharge Li ions are dissociated from the anode and migrate across the electrolyte and are inserted into the crystal structure of the cathode. At the same time the compensating electrons travel in the external circuit and are accepted by the host to balance the reaction. The process is completely reversible. Thus, the lithium ions pass back and forth between the electrodes during charging and discharging.

During charging, electrode and cell reactions can be represented as follows:

The positive electrode half reaction is

$$LiCoO_2 \rightleftarrows Li_{1-x}CoO_2 + xLi^+ + xe^-$$

The negative electrode half reaction is:

$$xLi^+ + xe^- + xC_6 \rightleftarrows xLiC_6$$

The net reaction during charging and discharging is represented as follows:

$$LiCoO_2 + C_6 \underset{\text{discharge}}{\overset{\text{charge}}{\rightleftharpoons}} Li_{1-x}CoO_2 + C_6L_x$$

Advantages

- They have high energy density than other rechargeable batteries.
- They are light weight.
- They produce high voltage out of 4 V.
- They have improved safety, i.e., more resistance to overcharge.
- No liquid electrolyte means they are immune from leaking.
- Fast charge and discharge rate.

Disadvantages

- They are expensive.
- They are not available in standard cell types.

Applications

- The Li-ion batteries are used in portable devices: these include mobile phones, laptops and tablets, digital cameras and camcorders, electronic cigarettes, handheld game consoles and torches (flashlights).
- Li-ion batteries are used in tools such as cordless drills, sanders, saws and a variety of garden equipment including whipper-snippers and hedge trimmers.
- Because of their light weight, Li-ion batteries are used for energy storage for many electric vehicles from electric cars to pedelec (**pedelec** is a bicycle where the rider's pedalling is assisted by a small electric motor), from hybrid vehicles to advanced electric wheelchairs, from radio-controlled models and model aircraft to the Mars Curiosity rover.
- They are used in cardiac pacemakers and other implantable devices.
- They are used in telecommunication equipment, instruments, portable radios and TVs, pagers.

2.7. FUEL CELLS

The conventional method of utilizing the chemical energy of a fuel involves several steps and can be depicted as follows:

Chemical energy ⟶ Thermal energy ⟶ Mechanical energy ⟶ Electrical energy.

Thermal energy liberated during the combustion of a fuel is used to convert water into steam. The steam is used to drive a turbine that drives the generator to produce electrical energy. There is a loss of energy in every step and only a maximum of 40% energy of the fuel is converted into electricity.

In a galvanic cell chemical energy is directly converted into electrical energy and the efficiency of conversion is very high. As in the case of combustion of fuels, the reaction taking place in a galvanic cell is also a redox reaction. But the only difference is that oxidation and reduction reactions in a cell take place at two different electrode surfaces separated from each other. The electrons liberated at the anode flow to the cathode, thereby generating electricity. Therefore, it was thought that if the oxidation of the fuel and reduction of the oxidant, oxygen, can be carried out separately at a distance between the two, then there can be an identical situation as in the case of a galvanic cell. Electrons liberated during the oxidation of the fuel can flow to the reduction zone of oxygen. The attempt in this direction by several scientists over quite a long period of time lead to the invention of fuel cells, in which chemical energy of a fuel is directly converted into electrical energy.

The principle of the fuel cell was discovered in 1839 by Sir William Grove, who has been acknowledged as the "Father of the Fuel Cell". Grove was interested in reversing the process of electrolysis, which is what a fuel cell precisely achieves. The term "fuel cell" was coined in 1889 by Ludwig Mond and Charles Langer, who attempted to use air and coal gas to generate electricity. In 1932, Francis Bacon improved on the platinum catalysts of Mond and Langer, and soon Harry Karl Ihrig, of Allis Chalmers Manufacturing Company demonstrated a 20 horsepower fuel cell powered tractor. NASA began using fuel cells in the late 1950s and continue to do so today. With the advent of new technologies, the applications of fuels cells have now been extended extensively in various fields.

Definition: A fuel cell is a galvanic cell in which the chemical energy contained in a readily available fuel oxidant system is converted directly into electrical energy by means of electrochemical processes in which the fuel is oxidized at the anode.

Like any other electrochemical cell, the fuel cell has two electrodes and an electrolyte. However, the fuel and the oxidizing agents are continuously and separately supplied to the two electrodes of the cell, at which they undergo reactions. These cells are capable of supplying current as long as they are supplied with the reactants.

A fuel cell essentially consists of the following arrangement:

Fuel / electrode / electrolyte / electrode / oxidant.

At the anode, fuel under goes oxidation:

$$\text{Fuel} \longrightarrow \text{Oxidation product} + ne^-$$

At the cathode, the oxidant gets reduced:

$$\text{Oxidant} + ne^- \longrightarrow \text{Reduction products}$$

The electrons liberated from the oxidation process at the anode can perform useful work when they pass through the external circuit to the cathode.

Advantages

Theoretically the efficiency can be 100%. But actually it is about 50–80%, owing to over potential and resistance of the cell.

- High efficiency of the energy conversion process.
- No moving parts and so elimination of wear and tear.
- Silent operation.
- Absence of harmful waste products.
- No need of charging.

Limitations

- Cost of power is high as a result of the cost of the electrodes.
- Fuels in the form of gases and oxygen need to be stored in tanks under high pressure.
- Power output is moderate.
- To have an appreciable voltage, a battery of fuel cells must be available.

Classification of Fuel cells

Fuel cells can be classified into three categories

(i) Direct fuel cells
(ii) Indirect fuel cells
(iii) Regenerative fuel cells.

In a **direct fuel cell**, the products of the cell reaction are discarded. In a **regenerative fuel cell**, the spent reactants are regenerated from the products by using thermal, electrical or photochemical methods. In an **indirect fuel cell**, the fuel of the cell is obtained from organic fuel. For example, in the reformer fuel cells organic fuels are converted into hydrogen. A biochemical fuel cell is another type of indirect cell in which a biochemical substance is decomposed by means of an enzyme in solution to produce hydrogen.

Hydrogen-Oxygen fuel cell

Hydrogen-oxygen fuel cell is a simplest type of fuel cell in which hydrogen gas is used as a fuel and oxygen gas as oxidant. A schematic diagram of $H_2 - O_2$ fuel cell is shown in Fig. 2.9.

The cell consists of a porous carbon electrode impregnated with catalysts such as finely divided platinum or palladium as anode. The cathode is also a porous carbon electrode impregnated with platinum or silver as catalyst. The electrolyte is an aqueous solution of KOH. The hydrogen gas fuel is continuously supplied at the anode and oxygen gas is supplied at the cathode. As the hydrogen gas diffuses through the anode, it is adsorbed on the electrode surface and reacts with hydroxyl ions to form water. At the cathode oxygen diffusing through the electrode is adsorbed and reduced to hydroxyl ions. These electrode reactions are summarized below:

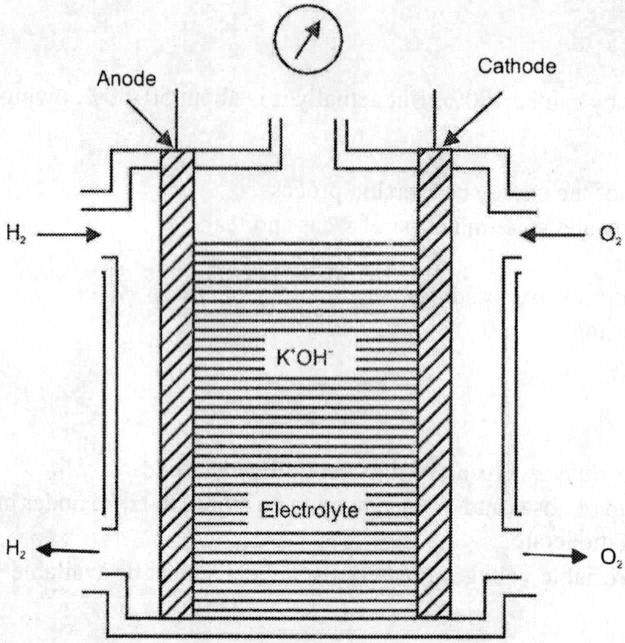

Fig. 2.9. H_2–O_2 Fuel cell

At the anode: $2H_2 \rightarrow 4H^+ + 4e^-$

$$4H^+ + 4OH^- \rightarrow 4H_2O$$

$$2H_2 + 4OH^- \rightarrow 4H_2O + 4e^-$$

At the cathode: $O_2 + 2H_2O + 4e^- \rightarrow 4OH^-$

The net cell reaction is, $2H_2 + O_2 \rightarrow 2H_2O$

The cell reaction is nothing but the combustion of hydrogen. It is to be noted that fuel cells do not store electric energy. They are merely energy conversion devices.

Methyl alcohol fuel cell

In a methanol fuel cell, methanol is used as the fuel and oxygen or air as oxidant. A typical methanol fuel cell is shown in Fig. 2.10.

Electrode systems

Electrodes are made up of platinum.

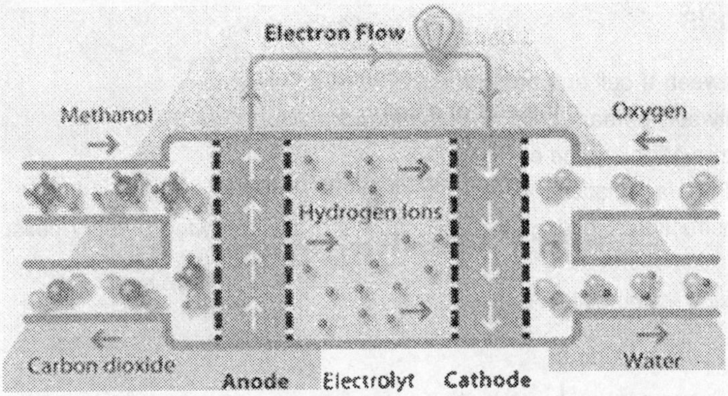

Fig. 2.10. Methanol fuel cell

Electrolyte

Sulphuric acid is used.

Electrode reactions

At the anode: $CH_3OH + H_2O \rightarrow CO_2 + 6H^+ + 6e^-$

At the cathode: $3/2\ O_2 + 6H^+ + 6e^- \rightarrow 3H_2O$

Fuel cell reaction: $CH_3OH + 3/2\ O_2 \rightarrow CO_2 + 2H_2O$

Fuel (methanol) and air or oxygen are fed to the electrodes. The cell potential is 1.21 V at 25 °C. The acid electrolyte offers the advantage of easy removal of CO_2, a product of the cell reaction.

Uses

In all kinds of portable, automotive and mobile applications like,

- Powering laptop, computers, cellular phones, digital cameras.
- Fuel cell vehicles (FCVs).
- Spacecraft applications.
- Any consumables which require long lasting power compared to Li-ion batteries.

REVIEW QUESTIONS

1. Differentiate between a cell and battery.
2. Differentiate between primary cells and secondary cells.
3. Explain the factors affecting the emf of a cell
4. Explain the cell characteristics.
5. Explain the construction, working and applications of the following primary cells.
   (a) Dry cell
   (b) Mercuric oxide – zinc cell
   (c) Silver oxide – zinc cell
   (d) Zinc– air cell
   (e) Nickel – metal hydride cell
6. Explain the construction of lead storage battery. Write the discharging and charging reactions.
7. Write a note on nickel – cadmium battery.
8. Define a fuel cell. How is it different from galvanic cell?
9. What are the advantages and limitations of fuel cells?
10. Explain the construction and working of the following fuel cells:
    (a) $H_2 - O_2$ fuel cell
    (b) Methanol fuel cell
11. Justify the following statements:
    (a) Dry cell can not be recharged.
    (b) In mercuric oxide – zinc cell, the electrolyte KOH is invariant.
    (c) The emf of a lead storage cell is dependent on the concentration of sulfuric acid.
    (d) Fuel cells are more efficient means to use the fuel energy than conventional methods.
    (e) Fuels cells are only energy conversion devices and not energy storage devices.
12. What are lithium ion batteries?
13. Explain the construction and working of a lithium ion battery.
14. What are the advantages and disadvantages of lithium ion batteries?
15. Enumerate the applications of lithium ion batteries.

CHAPTER 3

Corrosion and its Control

3.1. INTRODUCTION

A vexing problem associated with every use of metal object is corrosion. Silver tarnishes, copper develops platina and iron rusts. Corrosion occurs with all metals except the least active noble metals such as gold and platinum.

Rusting of iron and steel is the most familiar and outstanding corrosion process. It is the largest problem encountered throughout the world, mainly due to the fact that iron and steel are the most abundantly used materials. It is estimated that about 20% of iron produced annually is used just to replace iron objects that have been discarded due to rust damage.

The corrosion is a slow process taking place mainly on the metal surfaces, but the losses incurred due to corrosion are of high order. The losses can not be estimated by considering only the metal loss. The indirect losses are much higher. The fabrication cost and the cost incurred in preventing the corrosion are also have to be taken into consideration. The corrosion in metal objects like equipments, instruments, chemical plants, structures, etc., may make them inefficient, ineffective, useless and their unobserved defects due to corrosion may cause their failure. In such cases, the indirect losses may be much higher than the direct losses. When a structure like building or bridge collapses due to corrosion problems the loss also includes the loss of human life and properties, and the cost of subsequent repair and alternate arrangement. In some forms of corrosion, the corrosion is quite invisible and seen only when an accident takes place.

Corrosion once thought a problem concerned with only metallurgists and chemists is now a field of interdisciplinary study as metals are universal in their use. There cannot be any field of material use which does not use metals. May it be a small metal object like pin or a huge metallic structure like bridge or offshore structure, corrosion impairs their proper function and use. Therefore, corrosion has been correctly referred to as "cancer" to metals.

The knowledge of corrosion is, therefore, of great importance to the scientific community in general, and engineers in particular. It is the engineer who plans, designs and constructs the final metal product for use. In the process, due importance need to be given to the corrosion aspect. This can be successfully done only if he knows the mechanism of corrosion and the possible methods and mechanisms of control. This chapter highlights some of the important facts about corrosion and its control.

3.2. DEFINITION

Corrosion is defined as "the destruction or deterioration and consequent loss of metals through chemical or electrochemical attack by the environment". The process of corrosion in reality is the transformation of pure metal into its undesired metallic compounds. The life of a metal object gets shortened by the corrosion process. Corrosion can be a fast or slow process depending upon the metal, and the environment in which it is undergoing corrosion. The familiar examples of corrosion include:

(i) Rusting of iron – a reddish brown scale formation on iron and steel objects. It is due to the formation of hydrated ferric oxide.

(ii) Green scales formed on copper vessels. It is due to the formation of basic cupric carbonate. ($CuCO_3 + Cu(OH)_2$)

Why do metals undergo corrosion?

Most metals, except the noble metals, occur in nature in the form of their compounds such as oxides, sulfides, carbonates, chlorides, etc. The metals are extracted from these ores by reduction. The extraction of metals from their ores is an endothermic process, energy being supplied in the form of heat or electrical energy. Consequently, pure metals are relatively at higher energy state compared to their corresponding ores, and they have a natural tendency to revert back to their combined state. Therefore, when metals are put to use in various forms, they combine with the constituents of the environment and get converted into their compounds. Thus corrosion of metals can be considered as extractive metallurgy in reverse.

Corrosion of metal occurs either by direct chemical attack or by electrochemical attack on the metal by the corrosive environment. Direct chemical attack takes place generally in the absence of moisture. This type of corrosion is known as dry corrosion in which the constituents of the corrosion medium directly combine with the metal due to their chemical affinity to the metal. Most of the corrosion cases are electrochemical in nature, taking place by the electrochemical attack on the metal in the presence of moisture or a conducting medium. Such corrosion is known as wet corrosion.

Corrosion damage

- Poor appearance
- Maintenance and operating costs
- Plant shutdowns
- Contamination of product
- Loss of valuable products due to leakage
- Effects on safety and reliability in handling hazardous materials.
- Product liability

.3. DIRECT CHEMICAL CORROSION OR DRY CORROSION

Direct chemical corrosion occurs due to the direct chemical reaction between the metal and the gases present in the corrosion environment. This type of corrosion is generally observed in the absence of moisture or a conducting electrolyte medium, and therefore, known as **dry corrosion**. The term **scaling** or **tarnishing** are sometimes used to describe dry corrosion. Direct oxidation of metals and alloys on exposure to oxygen in air is a common example of chemical corrosion. Apart from oxygen, many corrosive gases such as HCl, H_2S, SO_2, Cl_2, F_2, NH_3, etc., also react with the metal and alloy surfaces when they come in contact with them. Two major factors deciding the rate of corrosion are the chemical affinity between the metal and the gas, and the nature of the corrosion product formed.

When a metal is exposed to a gas, it adsorbs the gas and slowly reacts with it and forms the corrosion product. For example, in oxidation corrosion, metal oxide is formed. The corrosion product so formed is deposited on the metal surface. Rate of further corrosion depends largely on the nature of the deposited corrosion product; whether it acts as a protective film or not. In this regard, the relative volumes of oxide and metal play an important part.

The protective and non-protective nature of the oxide layer can be predicted on the basis of **Pilling-Bedworth rule.** According to this rule an oxide is protective or non-porous, if the volume of the oxide is at least as great as the volume of the metal from which it is formed, and the oxide layer is porous or non-protective, if the volume of the oxide is less than the volume of the metal from which it is formed. When the film is non-porous, it acts as a barrier between the underlying metal surface and the corrosion environment, retarding / preventing further corrosion. For example, heavy metals such as aluminium, chromium, lead, tin, etc., form a protective oxide film.

Metals like alkali and alkaline earth metals form porous oxide films, which allow further diffusion of oxygen to come in contact with the fresh metal surface, thereby acting as a non protective film.

The criteria established by Pilling and Bedworth may be employed only for approximate evaluations of the stability of metals against oxidation corrosion. This is because, the resistance of the metal to oxidation depends not only on the continuity of the film, but also on many other factors. The oxide must possess:

- a co-efficient of expansion nearly equal to that of the metal substrate
- good adherence
- a high melting point
- a low vapor pressure
- good plasticity to resist fracture
- low electrical conductivity
- volume ratio close to (little higher) 1 to avoid compressive stresses or lack of surface coverage.

The extent of chemical corrosion by other gases such as CO_2, SO_2, Cl_2, H_2S, F_2, etc., also depends upon the formation of protective film or non protective film by the corrosion product. For example, the AgCl film, from the attack of chlorine on the silver metal surface, is non-porous and protective. $SnCl_4$ formed on the surface of tin by the attack of chlorine is volatile, and leaves the fresh tin surfaces for further attack.

3.4. ELECTROCHEMICAL THEORY OF CORROSION

Most of the corrosion problems are best explained on the basis of electrochemical corrosion. These are often called wet corrosion since aqueous medium or moist air is required for corrosion to take place.

According to electrochemical theory, corrosion of metals takes place due to the formation of anodic and cathodic regions on the same metal surface or when two different metals are in contact with each other in the presence of a conducting medium. At the anodic region oxidation reaction takes place and the metal gets converted into its ions, liberating electrons. Consequently, metal undergoes corrosion at the anodic region. At the cathodic region, reduction reaction takes place. Since the metal can not be reduced further, metal atoms at the cathodic region are unaffected by the cathodic reaction. Some constituents of the corrosion medium take part in the cathodic reaction.

The electrons liberated at the anodic region migrate to the cathodic region constituting **corrosion current.** The metal ions liberated at the anode and some anions formed at the cathode diffuse towards each other through the conducting medium and form a corrosion product some where between the anode and the cathode. Corrosion of metal continues as long as both anodic and cathodic reactions take place simultaneously. There can not be anodic reaction without cathodic reaction and vice versa.

Corrosion reactions

Anodic reaction is a simple oxidation reaction in which the metal atoms are converted into their ions liberating electrons.

At the anodic region, $M \rightarrow M^{n+} + ne^-$

For example, when iron undergoes corrosion,

$$Fe \rightarrow Fe^{2+} + 2e^-$$

But the cathodic reactions are more complicated than anodic reaction as the constituents of the corrosion medium are involved in the reaction, and are dependent on the nature of the corrosion environment. The most common types of cathodic reactions are either liberation of hydrogen or absorption of oxygen. Correspondingly the corrosion types are known as **hydrogen type corrosion** and **oxygen type corrosion**.

At the cathodic region

(a) **Liberation of hydrogen** takes place in the absence of oxygen.

(i) In acidic medium and in the absence of oxygen, the cathodic reaction is,

$$2H^+ + 2e^- \rightarrow H_2 \uparrow$$

(ii) In neutral or alkaline medium and in the absence of oxygen, hydroxide ions are formed with simultaneous liberation of hydrogen

$$2H_2O + 2e^- \rightarrow 2OH^- + H_2 \uparrow$$

(b) **Absorption of oxygen**takes place in the presence of oxygen.

(i) In acidic medium and in the presence of oxygen.

$$4H^+ + O_2 + 4e^- \rightarrow 2H_2O$$

(ii) In neutral or alkaline medium and in the presence of oxygen, hydroxide ions are formed by the following reaction.

$$2H_2O + O_2 + 4e^- \rightarrow 4OH^-$$

Corrosion of iron produces Fe^{2+} ions and OH^- ions at the anode and cathode sites respectively. These ions diffuse towards each other. Since smaller Fe^{2+} ions diffuse more rapidly than OH^- ions, their combination occurs more commonly near the cathodic region to produce insoluble $Fe(OH)_2$. In an oxidizing environment, it is oxidized to ferric oxide and the yellow rust is hydrated ferric oxide.

$$2Fe^{2+} + 4OH^- \rightarrow 2Fe(OH)_2$$

$$4Fe[OH)_2 + O_2 + 2H_2O \rightarrow 2[Fe_2O_3.3H_2O] \text{ [Rust]}$$

In the presence of limited oxygen, ferrous hydroxide is converted into magnetic oxide of iron (Fe_3O_4) and is known as black rust.

$$3Fe(OH)_2 + \tfrac{1}{2} O_2 \rightarrow Fe_3O_4.\ 3H_2O \text{ [Black rust]}$$

Summary

- A number of tiny galvanic cells are formed due to the formation of separate anodic and cathodic regions on the metal.
- Corrosion takes place at the anode and electrons liberated at the anode are consumed at the cathode.
- While anodic area is corroded, cathodic area remains unaffected.
- High level of oxygen concentration in the corrosion environment promotes and accelerates cathodic reaction, which in turn accelerates corrosion rate.
- Anodic and cathodic areas are formed on the same metal surface due to the presence of metal impurities in the bulk metal, difference in oxygen concentration on the metal surface, cracks and crevices in surface films and residual stress.

3.5. FACTORS INFLUENCING THE CORROSION RATE

There are several factors which influence the rate of corrosion to a varying degree. Some of the important factors, affecting the corrosion rate are discussed below.

(a) Nature of the metal: The tendency of a metal to undergo corrosion is dependent on the nature of the metal. In general, the metals with lower electrode potential values are more reactive than the metals with higher electrode potential values. The more reactive metals are more susceptible for corrosion. Thus the tendency of a metal to undergo corrosion decreases with increase in electrode potential. For example, the so called active metals like K, Na, Mg, Zn, etc., with low electrode potential values are highly susceptible for corrosion The noble metals such as silver, gold, platinum, etc., with higher electrode potential values are less susceptible for corrosion. However, there are few exceptions for this general trend as some metals show the property of passivity.

(b) Difference in potential between anodic and cathodic region: Larger the potential difference between the anodic region and cathodic region of the corrosion cell, higher is the corrosion rate. When potential difference is more, higher corrosion current is produced and the free energy decrease accompanying the process is higher and the corrosion rate is also higher. Therefore when two different metals with large difference in their electrode potentials are in contact with each other, the more reactive metal undergoes corrosion very fast.

(c) Nature of the corrosion product: The nature of the corrosion product largely decides the rate of further corrosion. The corrosion product formed on the surface of the metal may or may not act as a protective film. If the corrosion product deposited is insoluble, stable, uniform and nonporous, it acts as a protective film preventing the further corrosion of metal. A thin, invisible, impervious, continuous film formed on the surface acts as a barrier between the fresh metal surface and the corrosion environment. On the other hand if the corrosion product is soluble, unstable, non uniform and porous, the corrosion continues unabated. In such cases, the fresh metal surface is continuously exposed to the corrosion environment and corrosion of the metal surface takes place continuously. When the corrosion product formed is either soluble in the medium or formed away from the anodic and cathodic sites, then also the corrosion product does not have any protective value.

In oxidizing environments, metals like aluminium, chromium, titanium, etc., are highly passive as their oxides as corrosion products form protective films on the metal surface, preventing further corrosion. Metals such as iron, zinc, magnesium, etc., do not form any protective film and are highly susceptible for continuous corrosion, when exposed to oxidising environments. The protective films formed by the so called passive metals are of such a self healing nature that, when ruptured they themselves get repaired on exposure to the corrosion environment.

(d) The ratio of anodic to cathodic area: The rate of corrosion is greatly influenced by the relative sizes of anodic and cathodic areas. If a metal has a small anodic area and large cathodic area (i.e., the ratio of anodic to cathodic area is small), then the corrosion is more intensive and faster is the corrosion rate at the anodic region. As this ratio decreases, the corrosion rate increases rapidly. This is explained as follows. At the anode, oxidation takes place and electrons are liberated. At the cathode, these electrons are consumed. When the anode is smaller and cathode region is large all the electrons liberated at the anode, are rapidly consumed at the cathode region. This process makes the anodic reaction to take place at its maximum rate, thus increasing the corrosion rate. If the cathode is smaller, the consumption of electrons will be slower and the corrosion reaction as a whole will be slower. For instance, a broken coating of tin on iron surface enhances corrosion rate of iron / steel. Tin is cathodic to iron and the small exposed part of iron acts as anodic region and rest of the large tin coated surface acts as cathodic region. Because of small ratio of anodic to cathodic area, the corrosion rate is very high. For the same reason iron bolts and nuts are not used in copper boilers.

(e) Hydrogen overvoltage: A metal with low hydrogen overvoltage on its surface is more susceptible for corrosion, when the cathodic reaction is hydrogen evolution type. With lower hydrogen overvoltage, hydrogen gas is liberated easily and thus the cathodic reaction rate is faster. This will make the anodic reaction also faster, thereby promoting overall corrosion reaction. When the hydrogen overvoltage on the metal surface is high, cathodic reaction is slower and the corrosion of the metal also becomes slower.

(f) Temperature: The rate of a chemical reaction, in general, increases with rise in temperature. Corrosion process (redox reaction) is one such chemical reaction. Therefore, the rate of corrosion increases as the temperature increases. Increase in temperature increases the conductance of the corrosion medium, which also contributes to the increase in corrosion rate. In the case of corrosion resistant passive metals, the rise in temperature decreases the passive range and there by increases the corrosion rate.

(g) pH: In general, lower the pH of the corrosion medium, higher is the corrosion rate. However, some metals like Al, Zn, etc., undergo fast corrosion in highly alkaline solution. The pH of the solutions also decides the type of cathodic reaction.

(h) Presence of oxidizing agents: The presence of oxidizing agents increases the corrosion rate of the metal. Even noble metals undergo corrosion in the presence of oxidizing agents.

(i) Polarisation at anodic and cathodic region: During the process of corrosion, the polarisation of anode or cathode decreases the corrosion rate substantially. Because of anodic polarisation, the tendency of anodic metal surface to undergo oxidation decreases. That is, the rate of dissolution of metal as metal ion is decreased. This is generally the result of an increase in the concentration of ions of the dissolved metal in the vicinity of the electrode (concentration polarisation), or due to the formation of a protective film covering the anode (anodic passivity). Because of anodic polarisation, the anode reaction becomes slower and over all corrosion rate decreases.

Cathode polarisation retards the cathodic reaction. This can be due to chemical polarisation of the cathode, i.e., hindering the combination of cathode reactant with the electron. The retarded movement of the cathodic reactant to the cathode surface or retarded removal of the cathodic reaction product from the surface of the cathode also makes the cathodic reaction slower.

For the corrosion to continue, both anodic and cathodic reaction should take place simultaneously. The slow rate of one of the reactions makes the corrosion reaction slower.

The presence of depolarisers reduces the polarisation effect. In some instances, the addition of complexing agent can chelate the ions of the dissolving metal, reduce their concentration almost to zero, and consequently decrease the anodic polarisation markedly. As a result, corrosion processes are accelerated considerably. Such complexing agents in these cases are termed as anodic depolarisers. Similarly oxidising agents act as cathodic depolarisers by absorbing electrons from the cathode surface. Oxygen itself can act as a cathodic depolariser. The increased corrosion of water drainage equipment by mine water is caused primarily by the high concentration of ferric ions, and partly by cupric ions which are good cathodic depolarisers and thus stimulate corrosion.

(j) Electrical conductivity of the corrosion medium: As the conductivity of the corrosion medium increases, the corrosion rate also increases. Higher the conductivity of the medium, faster the ions can migrate between the anodic and cathodic regions of the corrosion cell, in turn, faster will be the exchange of electrons at the electrode surfaces. This facilitate higher corrosion rate. Therefore, corrosion problem is more in the sea water than in fresh water.

(k) Humidity: In general, corrosion rate increase with increase in the humidity in the medium. The presence of moisture in the atmosphere provides the conducting medium for the formation of galvanic cell, facilitating corrosion of the metal.

(l) Presence of impurities: The presence of certain impurities in the atmosphere enhances the rate of metal corrosion. For example, when SO_2 is present as impurity in the atmosphere, it combines with moisture or rain water forming sulfuric acid. In the presence of an acid metals like iron are more susceptible for corrosion. Similarly, presence of HCl also results in the formation of its acid, HCl, which again provides a corrosive environment for the metals.

3.6. TYPES OF CORROSION

In general, a metal gets destroyed or deteriorated due to corrosion. Depending upon the conditions, metals undergo corrosion by different mechanisms, giving rise to different types of corrosion cases. Some of the types of corrosion cases observed are discussed below.

(a) Galvanic Corrosion (Differential metal corrosion): Galvanic corrosion occurs when two dissimilar metals (galvanic couples) are in contact with each other in a corrosive conducting medium. The two metals differ in their tendencies to undergo oxidation; the one with lower electrode potential or the more active metal acts as anode and the one with higher electrode potential acts as cathode. The potential difference between the two metals is the cause or driving force for corrosion. The anodic metal undergoes corrosion and the cathodic metal is generally un attacked.

The following reactions occur during galvanic corrosion:

At the anode: $M \rightarrow M^{n+} + ne^-$
(Less noble metal)

At the cathode: Depending on the nature of the corrosion environment the
(More noble metal) cathode reaction may be either hydrogen evolution or oxygen absorption.

(i) $2H^+ + 2e^- \rightarrow H_2\,(g)$

or (ii) $O_2 + 2H_2O + 4e^- \rightarrow 4OH^-$

Examples

- Steel screws/rivets in copper sheet
- Steel pipe connected to copper plumbing
- Lead antimony solder around copper wire.

The rate of galvanic corrosion depends upon factors such as potential difference between the anodic and cathodic regions, ratio of anodic area to cathodic area, environmental factors, tendency of the metal to exhibit passivity, etc.

Galvanic Series

Different metals have different tendencies to react or to corrode. The familiar electrochemical series lists various elements in the order of their standard electrode potentials. The electrochemical series has, however, limited utility in corrosion studies and a poor guide in predicting corrosion behavior of metals and alloys in real environments. The standard electrode potential is the measure of the tendency of the metal to undergo corrosion or otherwise when it is in contact with one molar solution of its ions. But the real corrosion environments are quite different. The electro chemical series does not take passivity of the metals into account. For example, chromium is less noble than iron in the electrochemical series. But chromium is a passive metal and its alloying with iron imparts excellent corrosion resistance. Similarly metals such as aluminium and titanium, despite being anodic to iron exhibit a high degree of corrosion resistance, under various environments, contrary to the prediction by electrochemical series. Further, electrochemical series lists E^o values only for pure elements. But in engineering applications many alloys are used under various environments. The corrosion behavior of such alloys can not be predicted by electrochemical series.

In order to overcome these limitations of electrochemical series in corrosion studies, a series called galvanic series was introduced. The corrosion studies on various metals and alloys were actually performed in various environments. Then the metals and alloys were arranged in the order of their corrosion tendencies or corrosion resistance. An arrangement of metals and alloys in the order of their corrosion resistance in the given environment is referred to as galvanic series.

One such galvanic series in sea water is given in Table 3.1.

In general, the position of metals and alloys in the galvanic series agrees closely with the constituent elements in the electrochemical series. Passivity influences galvanic corrosion behaviors. Thus some metals and alloys occupy two positions in the galvanic series, depending on whether they are active or passive.

Table 3.1. Galvanic series of some metals and alloys in sea water

Metals and alloys	
Magnesium Magnesium alloys Zinc Galvanised steel	Active end
Aluminium	
Cadmium	
Steel or iron Cast iron	
Chromium steel (13%) (active) 18–8 Stainless steel (active)	
Lead Tin	
Nickel (active)	
Brasses Copper Silicon bronze	
Silver solder Nickel (passive)	
Titanium 18–8 Stainless steel (passive) Silver	
Graphite Gold Platinum	Noble end

Another interesting feature of the galvanic series is the brackets shown in the table 3.1. The alloys grouped in these brackets are close together in their corrosion tendencies and potential generated by these when coupled together, is not great. Thus, in practical applications, there is little danger of galvanic corrosion when these metals or alloys are in contact with each other.

(b) Differential aeration corrosion: Differential aeration corrosion occurs when a metal surface is exposed to differential air concentrations or oxygen concentrations. The part of the metal exposed to higher oxygen concentration acts as cathodic region and the part of the metal exposed to lower oxygen concentration acts as anodic region. Consequently, the poorly oxygenated region undergoes corrosion.

At the anode (less O_2 concentration),

$$M \rightarrow M^{n+} + ne^- \text{ (oxidation)}$$

At the cathode (more O_2 concentration),

$$O_2 + 2H_2O + 4e^- \rightarrow 4OH^- \text{(Reduction)}.$$

Corrosion of metals arising as a result of the formation of a oxygen concentration cell due to the uneven supply of air on the metal surface is known as differential aeration corrosion.

Some of the common examples are:

(i) Part of the nail inside the wall, being exposed to lower oxygen concentration than the exposed part, undergoes corrosion.
(ii) Window rods inside the frame suffer corrosion but not the exposed regions.
(iii) Paper pins inside the paper gets corroded, and the exposed part is free from corrosion.
(iv) Metal under dirt, dust, scale or water undergoes corrosion.
(v) Partially buried pipe line in soil or submerged in water undergoes corrosion below the soil or water where as the exposed part is free from corrosion.

Water line corrosion: Water line corrosion is a case of differential aeration corrosion, more prevalent in cases such as ocean going ships, water storage steel tanks, etc., in which a portion of the metal is always under water (Fig. 3.1)

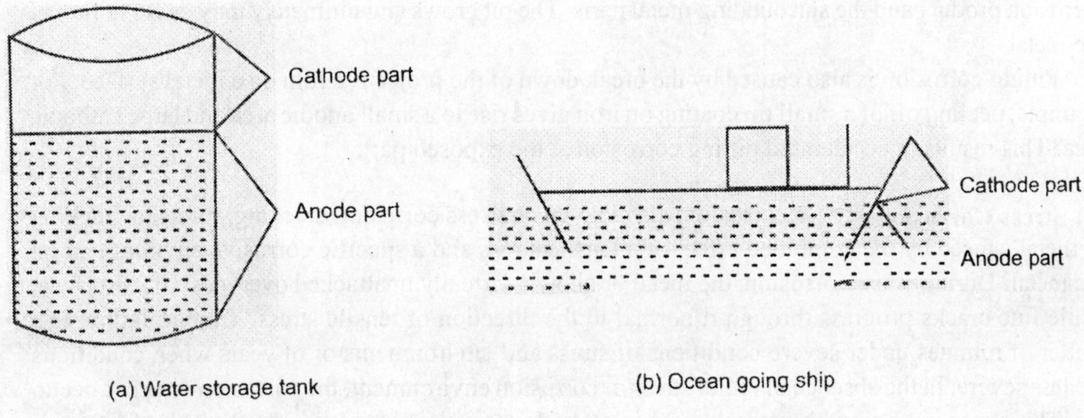

Fig. 3.1. Water line corrosion

The water line corrosion takes place due to the formation of differential oxygen concentration cells. The part of the metal below the water line is exposed only to dissolved oxygen while the part above the water is exposed to higher oxygen concentration of the atmosphere. Thus, part of the metal below the water acts as anode and undergoes corrosion and the part above the water line is free from corrosion. A distinct brown line is formed just below the water line due to the deposition of rust. Water line corrosion is common in water lifting pipes under water. Though ocean going ships suffer differential aeration corrosion, ships sunk under water for several years are free from corrosion. For the height of the ship sunk under water the difference in oxygen concentration between the top and bottom of the ship is negligibly small. Thus, the ship is exposed to almost uniform concentration of air and does not undergo differential aeration corrosion.

(c) Pitting Corrosion: Pitting corrosion is a localized and accelerated corrosion, resulting in the formation of pits or pin holes, around which the metal is relatively unattacked. Pitting is one of the most destructive forms of corrosion. It causes equipment to fail because of perforation with only a small percent weight loss of the entire structure. It is often difficult to detect pits because of their small size and also, because the pits are covered with corrosion products. Pitting is dangerous because it is a localized and intense form of corrosion and failures often occur with extreme suddenness. As such, it is rather difficult to assess precisely the life of metal component undergoing pitting corrosion. Loss of metal due to pitting is of little significance when compared to indirect losses such as production, repairs and replacement.

Pitting corrosion is characterized by small anodic area and large cathodic area, resulting in accelerated corrosion at the anodic area. It is an autocatalytic process, with the initially formed pit produces conditions which are both stimulating and necessary for the continuing activity of the pit.

The pitting corrosion is generally initiated by the deposition of extraneous matter such as sand, scale, water drop, dust, etc., or due to the break down of the protective film. The metal below the deposit is exposed to lower oxygen concentration, acts as anode and undergoes corrosion. The metal surrounding the deposit acts as cathode since it is exposed to higher concentration of oxygen. Because of the small anode and large cathode, accelerated corrosion takes place below the deposit. Once the corrosion product is formed, it further provides the condition for differential aeration below the corrosion product and the surrounding metal parts. The pit grows and ultimately may cause failure of the metal.

Pitting corrosion is also caused by the break down of the protective film on a metal surface. For example, peeling off of a small tin coating on iron gives rise to a small anodic area and large cathodic area. This results in accelerated pitting corrosion at the exposed part.

(d) Stress Corrosion: Stress corrosion, also known as stress corrosion cracking, refers to cracking of metal caused by the combined effect of a tensile stress and a specific corrosive environment on the metal. During stress corrosion, the metal or alloy is virtually unattacked over most of its surface, while fine cracks progress through it normal to the direction of tensile stress. Time to failure is a matter of minutes under severe conditions of stress and environments or of years when conditions are less severe. In the absence of either stress or corrosion environment, the corrosion does not occur.

The stress acting on the metal can be internal stress developed during the manufacture or fabrication of the articles, when the metals are subjected to mechanical operations such as pressing, hammering, bending, welding and riveting. The stress can also be external stress acting on the metal during service conditions. The metal atoms under stress are always at higher energy levels as compared to the ones free from stress. The stressed part of the metal therefore becomes more reactive than the stress free part. As a result, a corrosion cell is formed with the stressed part acting as anode and the stress free part acting as cathode. Under specific corrosion environments the stressed part undergoes corrosion, initiating the crack. In the presence of continuously acting stress and corrosion medium the crack propagates through the metal.

The corrosion medium is highly selective and specific in their action. For example, brass undergoes stress corrosion in the presence of ammonia and stainless steel in the presence of chloride ions and caustics.

Season cracking of brass and caustic embrittlement of steel are two classic cases of stress corrosion. Season cracking refers to the stress corrosion cracking of brass in the presence of ammonia.

Caustic embrittlement: Caustic embrittlement is a form of stress corrosion that takes place in boilers operating at high pressure, at the stressed regions of the boiler. Boiler plate under stress becomes brittle by the action of caustic alkali which causes failure of the boiler. Caustic embrittlement occurs at stressed parts of the boiler such as cracks, rivets, bends and joints.

The boiler feed water invariably contains some sodium carbonate (Na_2CO_3) left behind during the softening process. At high pressure and temperature, the residual Na_2CO_3 undergoes hydrolysis producing sodium hydroxide.

$$Na_2CO_3 + H_2O \rightarrow 2NaOH + CO_2$$

The caustic water seeps through the minute cracks, crevices between rivets and joints by capillary action. Inside the cracks and crevices water gets evaporated leaving behind NaOH. The concentration of alkali gradually increases on these sites due to poor circulation of water. When the concentration of NaOH reaches a value of 10%, it attacks the metals at the stressed region, dissolving it in the form of sodium ferroate (Na_2FeO_2). This sodium ferroate undergoes hydrolysis, depositing magnetite (Fe_3O_4) by either of the following reactions.

$$3Na_2FeO_2 + 4H_2O \rightarrow 6NaOH + Fe_3O_4 + H_2$$

Or $$6Na_2FeO_2 + 6H_2O + O_2 \rightarrow 12NaOH + 2Fe_3O_4$$

Sodium hydroxide is regenerated in the process. Thus the higher concentration of NaOH, once built up, continue to remain, maintaining the corrosion environment. The corrosion continues, developing cracks and making the metal brittle by the deposition of magnetite. The corrosion cell can be represented as follows:

Fe under stress (Anode)/Conc. NaOH/Dilute NaOH/Iron stress free (Cathode)

The iron in contact with concentrated alkali, being under stress, becomes anode, while stress free iron surrounded by dilute alkali becomes cathode. Corrosion takes place at the stressed region. The continuation of caustic embrittlement of boiler parts often results in boiler failure or boiler explosion.

Caustic embrittlement can be prevented by the addition of compounds such as sodium sulfate, tannin, lignin, phosphates, etc., which block the cracks or crevices, thereby preventing the infiltration of alkali.

3.7. CORROSION CONTROL

Corrosion of a metal is a natural spontaneous process, by which a metal is converted into a more stable compound state. Therefore, corrosion control is more realistic than corrosion prevention. The corrosion types are so numerous, the mechanisms of corrosion are so different, and conditions under which corrosion takes place are so varied that no single method can be used to control all possible corrosion cases. The choice of a control method depends on factors such as the type of the structure, the nature of prevailing corrosion, the residual stress in the fabricated articles, the nature of the environment and similar other factors. Some of the important methods commonly used in curtailing corrosion of metals are discussed in the following sections.

(a) Design and selection of materials

Selection of materials: The planning of corrosion prevention begins always with the selection of materials, i.e., the selection of suitable metal or alloy for a specific use under the prevailing conditions. Therefore, while selecting a metal or an alloy for the construction / fabrication of any structure / equipment, the following facts are to be taken into consideration.

- Noble metals such as platinum, gold, palladium, etc., do not undergo corrosion readily. But these metals cannot be used in most of the applications because of their prohibitive costs. Metals like iron, which are readily available and economically priced, are used extensively as construction metals.
- Pure metals show greater resistance to electrochemical corrosion than the impure ones. The presence of even a small impurity makes the metal heterogeneous and induces electrochemical corrosion. But pure metals may be more susceptible for direct attack from the corrosive media.
- Alloying of metals with suitable other metals improves the corrosion resistance. Steel is more corrosion resistant than pure iron. Addition of alloying metals such as chromium, nickel, molybdenum, etc., increases the corrosion resistance of steel due to the formation of self-healing protective oxide film on its surface, rendering the steel surface passive.
- The residual stress on any part of the metal causes stress corrosion. The fabricated metal article has to be relieved off any residual stress by proper heat treatment.
- The use of dissimilar metals, with large difference in their electrode potentials should be avoided. Otherwise more intense galvanic corrosion would result. However, under unavoidable circumstances, the metals chosen should be very close in their positions in the galvanic series.

Design: The most economical and elegant way of suppressing corrosion is through proper design of the metallic structure. A good design can minimize the corrosion in many metallic structures and equipment. It has been observed that the main cause for many premature failures of structures is due to the faulty design. A small modification in the design, however, may reduce the corrosion rate drastically. The following points are to be carefully taken into consideration while designing any metal structure or equipment.

- The sharp edges, sharp corners and sharp bends are the potential corrosion sites and must be replaced with round edges, round corners and round bends, respectively. At sharp edges and corners, protective coatings are usually thin and may not give the desired protection against corrosion. Further, at sharp bends, there is a sudden change in the direction of the fluid flow, resulting in erosion effect, which promotes corrosion.
- The design of a structure should be such that it does not retain fluid after draining off. Any stagnation of fluids leads to corrosion. The storage tanks, roofing sheets, etc., should be designed to avoid stagnation of water.
- The metal containers should not rest on the ground, but be raised above the ground level for uniform circulation of air to avoid differential aeration corrosion.
- The direct joining of two dissimilar metals should be avoided as this leads to galvanic corrosion. For example, the use of steel bolts and nuts in copper tanks, results in large cathodic area (copper tank) and small anodic area (steel bolts, nuts), leading to accelerated corrosion on bolts and nuts.
- When the use of two dissimilar metals in contact with each other is unavoidable, then the metals chosen should have a small difference in their electrode potentials or they should occupy closer positions in the galvanic series. Also, the smaller part should be of more nobler metal, in order to ensure smaller cathodic area and larger anodic area.
- The direct contact between the dissimilar metals should be avoided by insulating the parts from each other, thereby, avoiding the galvanic cell formation between them. Water repellant materials like rubber and plastic washers are to be used to avoid direct contact of dissimilar metals.
- The use of materials such as wood, cloth, paper, asbestos, etc., which absorb and retain moisture should be avoided in contact with the metals.
- Crevices, cracks, depressions, etc., should be avoided on the metal surface as they can collect dirt, dust, moisture, etc., which promote corrosion. The cracks, crevices, etc., should be filled with suitable fillers to avoid corrosion.
- Weld rather than rivet, tanks and containers as rivetted joints provide sites for crevice corrosion.

The general rule for design is "avoid heterogeneity". Hence, in design, attempt to make all conditions as uniform as possible throughout the entire system.

(b) Protective coatings

Application of protective coating is one of the important methods of corrosion control. The protective coating protects the metal from corrosion by acting as a barrier between the metal and the corrosion environment. The principal types of coatings applied on the metal surface are:

(i) Metal Coating
(ii) Inorganic Coating
(iii) Organic Coating

(i) Metal Coating: Deposition of a protective metal over the surface of a base metal is called metallic coating. It is a valuable and well tried method for improving the corrosion life of the underlying metal. Metal coatings are applied by electro deposition, flame spraying, cladding, hot dipping, diffusion coating and vapor deposition. From the corrosion point of view, metal coatings can be divided into two classes – **anodic coatings and cathodic coatings.**

Anodic Coatings: Anodic coatings are produced by coating a base metal with more active metals which are anodic to the base metal. For example, iron is coated with anodic and more active metals such as zinc, magnesium and aluminium. The one of the important characteristics of anodic coating is that, even if the coating is ruptured, the base metal does not undergo corrosion. The exposed surface of the metal is cathodic with respect to the coating metal and the coating metal preferentially undergoes corrosion. The protection is ensured as long as the anodic coating metal is still present on the surface. Therefore, anodic coating is also known as sacrificial coating. Galvanisation is a familiar example for anodic coating and is extensively used to protect iron and steel objects.

Galvanisation is a process of coating a base metal surface with zinc metal. Galvanisation of steel is favoured because of its low cost and eases of application. Galvanisation is carried out by hot dipping method. The galvanisation process involves the following steps:

- The metal surface is washed with organic solvents to remove organic matter on the surface.
- Rust and other deposits are removed by washing with dilute sulfuric acid.
- Finally the article is well washed with water and air dried.
- The article is then dipped in a bath of molten zinc, maintained at 425–430°C and covered with a flux of ammonium chloride to prevent the oxidation of molten zinc.
- The superfluous (excess) zinc on the surface is removed by passing through a pair of hot rollers, which wipes out excess of zinc coating and produces a thin coating.

Galvanisation is used extensively to protect iron from corrosion in the form of roofing sheets, fencing wire, barbed wire, buckets, bolts, nuts, nails, screws, pipes, tubes, etc. Galvanised steel reinforcements are used in building construction, where high degree of corrosion resistance is required.

Galvanised articles are however, not used for preparing and storing food stuffs, since zinc dissolves in dilute acids producing toxic zinc compounds.

Cathodic Coating: Cathodic coatings are produced by coating a base metal with a more noble metal which is cathodic. The metals such as copper, nickel, tin, silver are commonly used as cathodic coatings on steel. These coating metals are less reactive than the base metal and are less susceptible for corrosion. The cathodic coating provides protection only when it is undamaged, continuous and non porous. When the coating is discontinuous, an intense localized corrosion occurs due to the setting up of a galvanic cell that consists of a large cathodic area and a small anodic area. Tinning is the most commonly used cathodic coating on steel surfaces.

Tinning is a process of coating a base metal with tin. Tinning of iron or steel is carried out by hot dipping method and the process involves the following steps.

- The sheet is first washed with organic solvents to remove grease or oil deposits.
- Then treated with dilute sulfuric acid to remove rust and scale deposits. Finally it is washed well with water and air dried.

- The clean and dry sheet is passed through molten zinc chloride flux. The flux helps the molten metal to adhere on the metal surface.
- It is then passed through a tank that contains molten tin.
- Finally passed through a series of rollers immersed in palm oil. The rollers wipe out excess tin deposit and produce a continuous, thin coating of tin on the sheet. The oil prevents the oxidation of the tin coated surface.

Tin coating on steel provides complete protection against corrosion only if it covers the steel surface completely. If a small crack is formed in the tin coating, a galvanic cell is formed with exposed iron as anode and tin coating as cathode. An intense localized corrosion occurs at the exposed iron surface because of large cathodic area (tin) and small anodic area (iron).

The largest single use of tin is in coating of steel for manufacturing containers used for storing food stuffs such as jam, instant food, milk products, pickles, etc. Tin coated cans are also used for storing kerosene. Copper utensils are coated with tin to prevent contamination of food stuff with poisonous copper salts.

(ii) Inorganic Coating: Inorganic coatings are generally chemical conversion coatings. A surface layer of the metal is converted into a compound, by chemical or electrochemical reactions, which forms a barrier between the underlying metal surface and the corrosion environment. The chemical conversion coatings are different from other types of coatings in the sense that, they are the integral part of the metal itself. This types of coatings are formed on the metal surface by chemical dip, spray or by electrolytic methods. In addition to the corrosion resistance, this types of coatings also provide increased electrical insulation, and enhanced adherence for paints and other similar organic coatings.

Anodized Coating: Anodized coating is generally produced on non ferrous metals like Al, Zn, Mg and their alloys by anodic oxidation process, in which the base metal is made as anode, in an electrolytic bath of suitable composition, and by passing direct electric current. The most commonly used baths are chromic acid, sulfuric acid, phosphoric acid, oxalic acid, boric acid or their mixtures. Lead is generally used as the cathode. The anodic oxide film formed on Al in the bath of these materials tends to be porous and provides good adherence for paints and dyes. The strength and corrosion resistance of the anodized film can be increased by the so called sealing, which involves heating in boiling water or steam or metal salt solutions. The treatment changes porous alumina at the surface of coating into its monohydrate ($Al_2O_3 \cdot H_2O$), which occupies more volume, thereby the pores are sealed.

Anodized aluminium can also be produced in a number of colored finishes. Two types of methods used, are:

1. An organic dye is adsorbed on the freshly anodized surface and it is then sealed with boiling water.
2. A metal, often Ni, Co or Sn, is deposited into the pores. The colors resulting from this electro-coloring process are due to the interference effects and a wide range of tints can be produced by the variation of plating time and conditions.

Anodized articles are used as soap boxes, tiffin carriers, window frames, etc.

Phosphate Coating: Phosphate coating is obtained generally on steel surface by converting the surface metal atoms into their phosphates by chemical or electrochemical reactions. Chemical phosphating includes dipping and spraying and electrochemical phosphating includes cathodic and anodic phosphating.

The phosphating bath contains three essential components:

1. Free phosphoric acid
2. A primary metal phosphate like Fe, Mn and, Zn phosphates and
3. An accelerator such as nitrates, nitrites, chlorates, H_2O_2, etc.

The pH is usually within the range 1.8 and 3.2. Lower pH are favoured for immersion deposits while spray processes often employ a high pH. The reactions include dissolution of metal as ions, forming a phosphate with the bath solution which subsequently gets deposited on the surface of the metal.

Originally phosphate coating was applied only on iron and steel. But now it has been extended, to Zn, Al and their alloys.

The most important application of phosphate coating is as an under layer for paint finishes. Phosphating is a standard pretreatment before painting of automotive bodies and other steel sheet fabrications such as refrigerators and washing machines. In the case of car bodies, electrophoretic priming usually follows the phosphating stage. Phosphating is widely applied for the protection of metal parts which can not be painted or readily plated, such as nuts, bolts, the moving parts of machinery, etc. One of the important application is the phosphating of galvanized iron which is otherwise difficult to paint satisfactorily.

Organic Coatings: Coating of metal surfaces with organic materials, particularly paints and lacquers is the most widely used anticorrosion coatings and probably accounts for more than half of all costs spent on long term anticorrosion measures. The functions of organic coatings are of twofolds:

1. The coatings serve to keep out air and moisture from the metal surface or serve as a barrier between the metal surface and the corrosion environment.
2. The pigments, (red lead, zinc chromate, etc.) or drying oils (linseed oil, wood oils, etc.) present in the paint often exert an inhibitive action by electrochemical and other means.

The requirements of a good organic coating are:

- It should adhere tenaciously to the metal surface and improve its physical appearance.
- The film formed should be continuous, uniform and impervious to air and water.
- Should be chemically inert to the corrosion environment.
- Should have reasonably long life.
- Should have proper application methods.

The performance of the paint and lacquer coatings depends to a large extent on the application technique and even slight negligence at any stage may cause failure of the coatings. The application of coating involves the following sequence.

Surface preparation, priming, filling, sanding and finishing or top coats: Surface preparation includes degreasing the surface, and ensuring dry and scale free surface. A high degree of protection is attained by application of several coats of paints of different composition. The first coat is the **primer** such as phosphate coating which must be strongly adhered to the surface of the articles to be subsequently painted. **Fillers** such as nitrocellulose, epoxides, etc. are applied on the well dried primed surface in order to improve the external appearance of the paint and to produce a smooth coating. After drying, the roughness and irregularities on the surface are smoothened by means of emery paper (sanding) before applying the finishing or top coats.

Organic coatings are applied by different methods such as brushing, spraying, dipping, roller coating, etc. Break down of the organic coating may cause severe corrosion of the exposed metal parts.

(c) Corrosion Inhibitors

Corrosion inhibitors are substances, which when added in small concentrations to a corrosive environment decreases the corrosion rate. The inhibitors reduce the corrosion rate either by reducing the probability of its occurrence (deterrent) or by reducing the rate of attack (retardant) or by doing both. They are mostly used to provide protection to systems in which corrosion environment is recirculated or confined for a longer period such as internal combustion engine, recirculating cooling water systems and pipe lines for transporting oils, etc. The inhibitors provide protection against corrosion by retarding either anodic or cathodic reactions. On the basis of this property, they are broadly classified as

(i) Anodic inhibitors, and
(ii) Cathodic inhibitors

(i) Anodic Inhibitors: Oxidation of the metal is the reaction that occurs at anodic sites during corrosion. For example, in rusting of iron, Fe^{2+} ions are formed at the anodic region. If the formation of Fe^{2+} is prevented, the cathodic reaction also stops and thereby the corrosion process is retarded. This is achieved by the addition of anions such as chromate, tungstate, molybdate, phosphate, etc. These anions combine with metal ions formed at the anodic region, forming the sparingly soluble respective salts. These compounds formed are deposited on the anodic sites forming protective films, which act as barrier between the fresh metal surface and the corrosion medium, there by preventing further anodic reaction.

Anodic inhibitors are found to be effective only when sufficient amount of the inhibitor is added into the corrosion medium. When insufficient amount of the inhibitor is added, the situation is more disastrous than not adding at all. This is because of the fact that insufficient quantity of inhibitor can not cover the entire anodic surface with the protective film. The small anodic regions exposed to the corrosion medium undergo accelerated corrosion due to the smaller ratio of anodic area to cathodic area.

(ii) Cathodic Inhibitors: The two important types of cathodic reactions are liberation of hydrogen or absorption of oxygen. Therefore, there are two distinct approaches in achieving inhibition of cathodic reactions. That is, either by preventing/retarding the liberation of hydrogen or by preventing/ retarding the absorption of oxygen.

Inhibition of hydrogen liberation: The liberation of hydrogen at the cathode can be prevented either by preventing/retarding the diffusion of H^+ ions to the cathode or by increasing the hydrogen overvoltage on the metal surface.

The diffusion of H^+ ions to the cathode is prevented by the addition of certain organic compounds which contain nitrogen or sulfur. Urea, thiourea, aliphatic amines, mercaptans and heterocyclic compounds are widely used as cathodic inhibitors. Such substances, when added to the corroding environment, are adsorbed on the cathodic sites forming a protective film, preventing the H^+ ions from coming in contact with the cathodic metal surface.

The evolution of hydrogen at the cathode can also be prevented by increasing the hydrogen overvoltage. This is achieved by the addition of oxides of arsenic, antimony or salts like sodium meta arsenite. They deposit as adherent metallic film on the cathode region and thereby prevent the liberation of hydrogen, as the hydrogen over voltage on these metals is very high.

Inhibition of oxygen absorption: The absorption of oxygen can be stifled either by removing the oxygen from the corrosive media or by simply decreasing the diffusion rate of oxygen to cathode. The first objective is achieved by adding reducing agents or oxygen scavengers such as hydrazine, sodium sulfite, etc., into the corrosion medium. These substances remove oxygen from the corroding environment by reducing it.

$$N_2H_4 + O_2 \rightarrow N_2 + 2H_2O$$

$$2Na_2SO_3 + O_2 \rightarrow 2Na_2SO_4$$

The second objective is achieved by adding salts such as $ZnSO_4$, $MgSO_4$, $NiSO_4$, etc., into the corrosion medium. The cations of the salts migrate towards the cathode surface and react with the cathodically formed alkali to deposit their hydroxides on the cathodic sites.

$$Zn^{2+} + 2OH^- \rightarrow Zn(OH)_2$$

$$Mg^{2+} + 2OH^- \rightarrow Mg(OH)_2$$

The protective film being impermeable to oxygen, prevents its diffusion to the cathodic region. The action of an inhibitor depends on the nature of the metal to be protected as well as the corrosive environment. It is, therefore, necessary to choose an appropriate inhibitor for a particular system.

Limitations

- They contaminate the environment.
- Many of the inhibitors are toxic, can not be used in systems which come in contact with humans.
- Can be used only in closed systems in which the corrosive environment is either contained or recirculated.
- Generally lose their effectiveness rapidly as the concentration and temperature of the environment increase.

(d) Cathodic Protection

Cathodic protection is a method of protecting a metal or alloy from corrosion by converting it completely into cathodic and no part of it is allowed to act as anode.

Principle: Metals normally undergo corrosion by electrochemical process with the formation of anodic and cathodic regions in contact with each other. The corrosion of metal takes place at the anodic region whereas at the cathodic region, the metal is unaffected. Therefore, corrosion can be prevented by eliminating the anodic sites and converting the entire metal into cathodic area.

Cathodic protection can be achieved by the following methods:

(i) Sacrificial anode
(ii) Impressed current.

(i) Sacrificial Anode Method: In this method, the protected metal structure is converted into a cathode by connecting it to a more active (less noble) metal. This active metal acts as an auxiliary anode. Zinc, magnesium and aluminium are the common auxiliary anodes used in this method. These metals, being more active, act as anode and undergo preferential corrosion, protecting the metal structure. Since the anodic metals are sacrificed to protect the metal structure, the method is known as sacrificial anode method. Exhausted sacrificial anodes are replaced by new ones as and when required. Examples of sacrificial anode methods are:

- A magnesium block connected to a buried oil storage tank (Fig 3.2(a))
- Mg bars are fixed to the sides of ocean going ships to act as sacrificial anodes (Fig 3.2 (b))
- Mg blocks are connected to buried pipe lines (Fig. 3.2(c))

Advantages

- The method is simple.
- Low installation cost
- Minimum maintenance cost
- Does not require power supply

Disadvantage

- Involves recurring expenditure for replacement of consumed anodes.

(ii) Impressed current method: Another method of providing cathodic protection is by applying a direct current larger than the corrosion current. The protected metal is made cathodic by connecting it to the cathode of the external source of current. The anode of the source is connected to an inert electrode (Fig. 3.3).

The metal structure being cathode does not undergo corrosion. Anode being inert remains unaffected. Graphite is widely used as the inert anode in this method. Platinum, silicon, iron are also used as anodes.

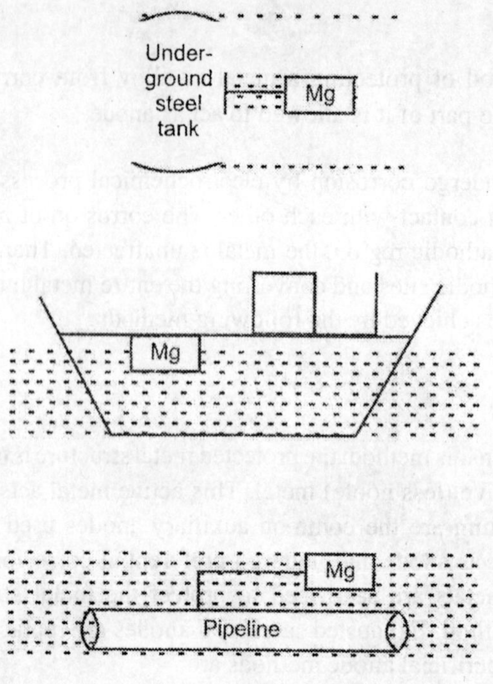

Fig. 3.2. Cathodic protection by sacrificial anode method

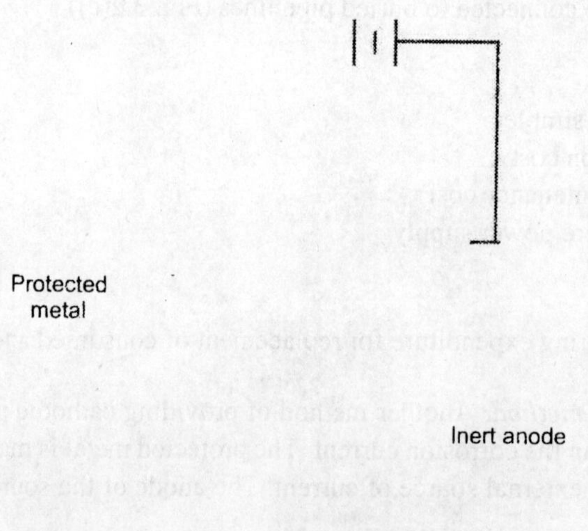

Fig. 3.3. Impressed current method

Advantages

- One installation can protect large area of metal
- Low maintenance cost

Limitations

- Rather expensive, since it needs high current for safe protection of structure.
- If the impressed current is not uniform on the entire surface of the protected structure, localized corrosion takes place on the protected metal.
- If hydrogen liberation is the cathodic reaction on the metal surface, then the protected metal may suffer hydrogen embrittlement.

(e) Anodic Protection

The application of anodic protection in preventing corrosion is a recent method. The prevention of corrosion by the impressed anodic current is called as **anodic protection.** The method, suggested by Edeleanu in 1954, is based on the formation of a protective film on metals by externally applied anodic current. Generally, when a metal is made as anode and current is applied, the dissolution rate of the metal tends to increase. But in the case of passive metals with active – passive transitions such ass nickel, iron, chromium, titanium and their alloys, application of suitable current makes them passivated and decreases the rate of dissolution of metals. The active – passive transition of these metals with the application of potentials can be explained from the potential – current curves as shown in the Fig. 3.4.

The curve is obtained by applying a known potential to the metal specimen and measuring the current changes. As the potential is increased, initially the current also increases (AB), indicating the dissolution of metal. This trend continues until the current reaches a critical value (i_{crit}) and passivation due to the development of oxide layer sets in. This potential is called passivating potential (Ep).

Above Ep, the current flow decreases and reaches a minimum value called passivation current, $i_{passive}$. The decrease in current is due to the formation of a passive film on the metal. If the potential

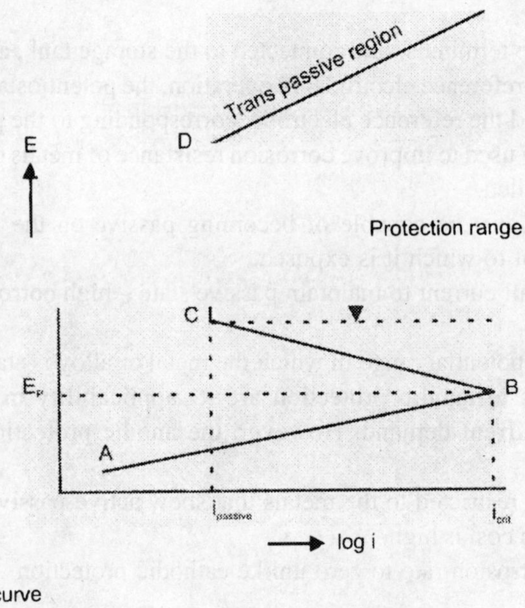

Fig. 3.4. Potential-current curve

is further increased, the metal remains unattacked up to a particular potential is reached (CD). In this range ($E_c - E_d$), corrosion rate of the metal is very small. This potential range, in which anodic protection can be achieved, is called protection range or passive region. The optimum potential for anodic protection is midway in the passive region, since it permits slight variation in the controlled potentials, without affecting corrosion rate. Beyond the point D, any further increase in potential tends to increase the current due to the dissolution of metal. It is called trans passive region.

The anodic protection to a structure is applied by using a device called potentiostat. It is an electronic device, that maintains a metal at a constant potential with respect to a reference electrode. The anodic protection of steel tank containing sulfuric acid is shown in Fig. 3.5.

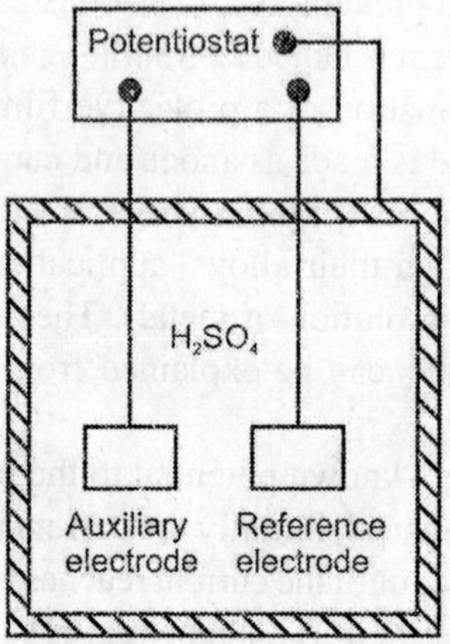

Fig. 3.5. Anodic protection of steel tank containing H_2SO_4

The potentiostat has three terminals, one connected to the storage tank, another to an auxiliary cathode (Pt) and the third to a reference electrode. In operation, the potentiostat maintains a constant potential between the tank and the reference electrode, corresponding to the passive range. Anodic protection can be successfully used to improve corrosion resistance of metals or alloys, provided the following conditions are fulfilled.

- The metal or alloy must be capable of becoming passive on the application of anodic current in the reagent to which it is exposed.
- It must require a small current to maintain passive state – high corrosion resistance at low energy consumption.
- A sufficiently broad potential range in which the metal or alloy is stably passive.

The primary advantages of anodic protection are its applicability in extremely corrosive environments and its low current demand. However, the anodic protection has the following limitations:

- Anodic protection is restricted to the metals that show active-passive behaviour.
- Its initial installation cost is high.
- It can not reduce corrosion rate to zero unlike cathodic protection.

REVIEW QUESTIONS

1. Define corrosion. Why do the metals show the tendency to undergo corrosion?
2. Explain electrochemical theory of corrosion.
3. Explain the effect of following factors on corrosion rate:
   (i) Nature of metal
   (ii) Nature of corrosion product
   (iii) The ratio of anodic area to cathodic area
   (iv) Hydrogen overvoltage
   (v) Polarisation at the electrode area.
4. Explain the following types of corrosion
   (i) Galvanic corrosion
   (ii) Pitting corrosion
   (iii) Differential aeration corrosion
   (iv) Stress corrosion
5. Write notes on the following:
   (i) Water line corrosion
   (ii) Galvanic series
   (iii) Caustic embrittlement.
6. Explain the following methods of corrosion control:
   (i) Cathodic protection
   (ii) Anodic protection
   (iii) By using inhibitors.
7. What are the types of protective coatings? How do protective coatings help in controlling corrosion?
8. Differentiate between anodic coatings and cathodic coatings.
9. Write notes on:
   (i) Phosphate coating
   (ii) Anodization
10. What are the differences between sacrificial anode method and impressed current method of cathodic protection? What are the limitations of impressed current method?
11. Account for the following:
    (i) Zinc in contact with silver undergoes corrosion faster than zinc in contact with copper.
    (ii) Though aluminium is with lower electrode potential than iron it has higher corrosion resistance than iron in oxidizing environments.
    (iii) Iron bolts in copper vessel are undesirable.
    (iv) Ruptured tin coating on iron is more disastrous than not having the coating at all.
    (v) Even if the zinc coating on iron is discontinuous, iron is free from corrosion.
    (vi) Inorganic coatings are generally chemical conversion coatings.
    (vii) Use of insufficient quantity of anodic inhibitors is more disastrous than not using it at all.
    (viii) Part of the nail inside the frame undergoes corrosion but the exposed part does not.
    (ix) Ocean going ships suffer differential aeration corrosion, but ships sunk under sea do not.
    (x) Metal under water drop undergoes accelerated corrosion.
    (xi) Anodic coatings are sacrificial coatings
    (xii) Galvanic corrosion is better explained by galvanic series than electrochemical series.
    (xiii) Whenever two different metals are used in contact with each other, the smaller part should be of nobler metals.
    (xiv) Welding is better than rivetting to join metal parts.
    (xv) Two different metals in contact with each other should be selected in such a way that their positions in the galvanic series should be close to each other.
12. Explain the mechanism of inhibitor action in corrosion control.
13. Explain direct chemical corrosion.

CHAPTER 4

Metal Finishing

4.1. INTRODUCTION

In majority of engineering applications of metals, the need is for an ideal metal or alloy which can meet several requirements such as corrosion resistance, wear resistance, mechanical properties, electrical and thermal properties, ease of fabrication, etc. However, no single metal can have all these ideal properties. Certain properties like corrosion resistance, demands the use of expensive noble metals and alloys. But the prohibitive cost and the difficulties in the fabrication of these metals compel the use of easily available, relatively low priced materials such as steel. Therefore, in general, a metal or alloy used for a purpose does not meet all the requirements. In such cases, provisions have to be made to improve their lacking properties. Metal finishing is one of the methods employed for the purpose.

Metal finishing is the name given to a wide range of processes carried out in order to modify the surface properties of a metal. These processes involve deposition of a layer of another metal or a polymer, conversion of a surface layer of atoms into oxide film or any other inorganic compound film and any such process, which ultimately modify the surface properties of the metal.

The large numbers of metal articles we use in our daily life have undergone a metal finishing process. Metal finishing, which was originally introduced as a decorative finish to enhance the values of metal articles by improving their appearance, has since been developed to such an extent that there is hardly any industry which does not make use of this process.

4.2. TECHNOLOGICAL IMPORTANCE OF METAL FINISHING

Metal finishing finds extensive applications in variety of industries, extending to many electronics, engineering and metal processing companies as well as to both large and small specialist firms. The technological importance of metal finishing is in imparting certain additional properties to the materials in addition to their intrinsic properties so that the utility of the materials is increased. These include

- Imparting higher corrosion resistance.
- Imparting improved wear resistance.
- Providing electrical and thermal conducting surface.
- Imparting thermal resistance.

- Imparting hardness.
- Building up of material or restoration.
- Providing optical or thermal reflectivity.
- Imparting improved solderability.
- Manufacturing electrical and electronic components such as printed circuit boards, capacitors, contacts, etc.
- Electroforming of articles.
- Electrochemical machining, electro-polishing and electrochemical etching.

Five important techniques of metal finishing are

(i) Electroplating of metals or alloys
(ii) Electroless plating of metals or alloys
(iii) Immersion plating of metals
(iv) Chemical conversion coating
(v) Electrophoretic painting

4.3. ELECTROPLATING

The term electroplating refers to the deposition of a metal on to the surface of another metal, alloy or any conductor in general, by the process of electrolysis. It consists of a process, by which a metal is deposited on another metal or alloy by passing a direct current through an electrolyte solution containing the metal ions to be deposited.

Therefore, electroplating can be defined as "a process in which a base metal is coated with a thin and uniform layer of another suitable metal by electrolytic deposition".

The common coating metals used are Zn, Cu, Ni, Cr, Ag, Pt, Au, etc. Electrodeposition is an important and frequently used method in industries for producing metal coating.

Since the process of electroplating essentially involves electrolysis, the three important factors governing the process of electrolysis, namely, **polarization, decomposition potential** and **overvoltage** govern the electroplating processes also.

4.4. POLARIZATION, DECOMPOSITION POTENTIAL AND OVERVOLTAGE

(i) Polarization: In an electrochemical cell, the electrical energy is produced due to a decrease in free energy of a spontaneous redox reaction that takes place within the cell. Under reversible conditions, the electrical energy available is exactly equal to the free energy decrease. Under irreversible conditions the electrical energy available is less than the free energy decrease.

Electrolysis involves conversion of electrical energy into chemical energy, i.e., chemical changes are brought about by the expense of electrical energy. Since electrolysis is a nonspontaneous process, the minimum energy required to carry out such a process is equal to the free energy increase accompanying the change, and this is equal but opposite in sign to the free energy decrease accompanying the reverse spontaneous process. This is true when electrolysis is carried out reversibly. However, in irreversible conditions, the potential to be applied for electrolysis is higher than the reversible emf of the cell. A cell that requires excess voltage over the theoretical is said to be **polarized.** The excess voltage is called **polarization voltage** and the phenomenon is referred to as **polarization.**

The polarization of an electrolytic cell arises when both or one of the two electrodes undergo polarization. The observed polarization can be of two types, namely, **(a) concentration polarization and (b) overvoltage.**

Concentration polarization results from the changes in concentration that occurs in the vicinity of an electrode during the current flow. If the depletion of concentration of the ions near the electrode is not compensated by diffusion of ions from the bulk of the solution, a concentration gradient is set up, developing a cell which can develop an emf opposing the applied emf. On the other hand overvoltage is a polarization potential whose source lies in a process at the electrode which takes place irreversibly and is thus a phenomenon intimately associated with the nature of the electrode and the processes occurring at its surface. The total polarization observed is the sum of both effects. However, concentration polarization can be eliminated by stirring the solution and in such conditions polarization voltage observed is essentially due to overvoltage.

(ii) Electrolysis and Decomposition Potential: Generally an electrolytic cell consists of a single electrolyte into which a pair of electrodes is dipped. For example, in the electrolysis of water, a pair of platinum electrodes is immersed in a solution of an acid or base. It is found experimentally that a potential of about 1. 7 V must be applied to the cell before there sets in a continuous evolution of hydrogen and oxygen. When a potential of less than 1.7 V is applied, the observation is a momentary surge of current which rapidly falls to zero. The reason for this is that, due to the initial passage of current, small amounts of hydrogen and oxygen are liberated at the cathode and the anode respectively. These gases are adsorbed on the cathode and the anode respectively and the inert platinum electrodes are converted into active hydrogen and oxygen gas electrodes. The secondary cell formed produces a back emf opposing the applied emf. At potentials below 1.7 V, the back emf nullifies the effect of applied potential, and continuous electrolysis takes place only when the applied emf is sufficient enough to overcome the back emf.

Since, even inert electrodes are converted into active electrodes on electrolysis, and produce a back emf, it is necessary to apply a minimum required potential to produce continuous electrolysis of an electrolyte. This minimum voltage need to be applied in order to have continuous electrolysis of an electrolyte is called the **decomposition potential**.

The arrangement for the determination of decomposition potential of an electrolyte is illustrated in Fig. 4.1. By means of the variable resistance C it is possible to regulate the applied voltage and this in turn can be measured by the voltmeter. The milliammeter is used to measure the current passing through the cell. To measure the decomposition potential, a series of current vs applied voltage readings are taken. Up to a definite potential, the current flow is very small and thereafter it increases with increase in potential. A plot of current versus applied potential is as shown in Fig 4.2. Intersection of the two straight line portions gives decomposition potential.

It is expected that the decomposition potential of any solution is approximately equal to the reversible emf of the galvanic cell consisting of the products liberated at the electrodes in contact with the same solution. But in some cases, the decomposition potential is considerably greater than the calculated reversible decomposition potential. For example, theoretically the decomposition

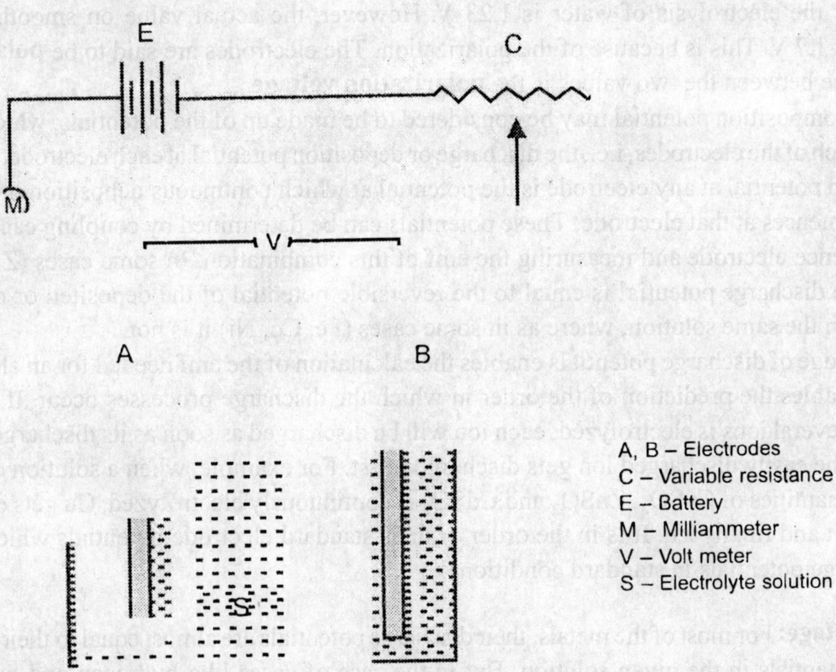

Fig. 4.1. Determination of decomposition potential

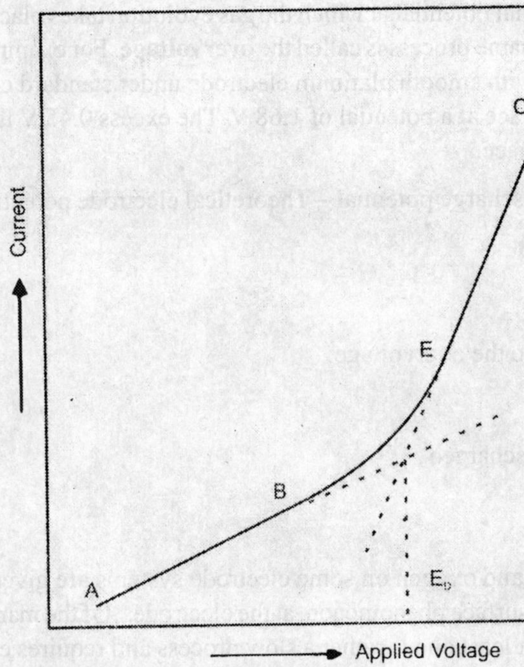

Fig. 4.2. Decomposition potential

potential for the electrolysis of water is 1.23 V. However, the actual value on smooth platinum electrodes is 1.7 V. This is because of the polarization. The electrodes are said to be **polarized** and the difference between the two values is the **polarization voltage**.

The decomposition potential may be considered to be made up of the potentials, which must be applied to each of the electrodes, i.e., the discharge or deposition potential at each electrode. Discharge or deposition potential at any electrode is the potential at which continuous deposition or discharge of ions commences at that electrode. These potentials can be determined by coupling each half-cell with a reference electrode and measuring the emf of this combination. In some cases (Zn, Cd, Cu, Cl_2, Br_2) the discharge potential is equal to the reversible potential of the deposited or discharged metal / gas in the same solution, where as in some cases (Fe, Co, Ni) it is not.

Knowledge of discharge potentials enables the calculation of the emf needed for an electrolysis. This also enables the prediction of the order in which the discharge processes occur. If a solution containing several ions is electrolyzed, each ion will be discharged as soon as its discharge potential is reached; the easily discharged ion gets discharged first. For example, when a solution containing equimolar quantities of $CuSO_4$, $ZnSO_4$ and $CdSO_4$ is continuously electrolyzed, Cu gets discharged first, Cd next and finally Zn. It is in the order of their standard electrode potentials which are also their discharge potentials in standard conditions.

(iii) Overvoltage: For most of the metals, their discharge potentials are almost equal to their reversible electrode potentials in the given solution. But in the case of gases like hydrogen and oxygen, the discharge potentials are appreciably higher than their reversible electrode potentials. The excess voltage is referred to as over voltage of the gas.

The difference between the actual potential at which the gas evolution takes place at the electrode and the theoretical potential for the same process is called the **overvoltage**. For example, the reversible electrode potential of oxygen gas with smooth platinum electrode under standard conditions is 1.23 V. But actual gas evolution takes place at a potential of 1.68 V. The excess 0.45 V is the overvoltage of oxygen on smooth platinum surface.

$$\text{Overvoltage of oxygen} = \text{Actual discharge potential} - \text{Theoretical electrode potential}$$

$$= 1.68 - 1.23$$

$$= 0.45\text{V}$$

The following factors influence the overvoltage:

(i) Electrode surface
(ii) Nature of the electrolyte
(iii) Nature of the substance discharged
(iv) Current density
(v) Temperature

The overvoltages of hydrogen and oxygen on some electrode systems are given in Table 4.1.

Overvoltage is believed to be a surface phenomenon, at the electrodes. Of the many steps involved in gas evolution at the electrodes, at least one is rather a slow process and requires extra energy. The extra energy is supplied by the applied voltage and this extra energy is closely related to overvoltage.

Table 4.1. Overvoltages of hydrogen and oxygen

Electrode	*H_2 – overvoltage (V)*	*O_2 – overvoltage (V)*
Platinized Pt	0.00	0.25
Iron	0.08	0.24
Platinum	0.09	0.45
Nickel	0.22	0.06
Cadmium	0.48	0.43
Lead	0.64	0.32
Zinc	0.70	
Mercury	0.80	

Overvoltage plays a role in many electrode proccsses which are technically important. The electrodeposition of most metals from solution takes place with very little overvoltage and hence at essentially their reversible potentials. But in aqueous solutions hydrogen ions are also present, and consequently two cathodic reactions are possible as both metal ions and hydrogen ions compete for getting discharged. If the potential required for metal deposition is less than that for hydrogen, product of electrolysis at the cathode will be the metal and if the reverse is true liberation of hydrogen will occur.

If the metals are compared with hydrogen, the metals with higher standard electrode potentials than hydrogen will preferentially be get deposited before the evolution of hydrogen. But, for the metals with lower standard electrode potential than hydrogen, we expect that the deposition of these metals is possible only after the liberation of hydrogen is completed. Contrary to these expectations, it is observed that, many metals with negative electrode potentials can be deposited from acidic solutions. For example, if zinc to be electrodeposited from a solution of 1M $ZnSO_4$ solution at pH = 4, the discharge potential for zinc is –0.76V. Under the same conditions with an over-voltage of 0.70V on Zn, the discharge potential for hydrogen is –0.236–0.70 = –0.936V. Since the value is more negative than that for Zn, it is zinc that is deposited first. Similarly some other metals can also be deposited from acidic solutions because of the hydrogen overvoltage on these metal surfaces.

Hydrogen overvoltage indicates the measure of the tendency of hydrogen to get liberated on a metal surface. Higher the hydrogen overvoltage, more difficult is the liberation of hydrogen on that metal surface.

4.5. THEORY OF ELECTROPLATING

Electroplating is essentially an electrolysis process. Metal ions in the electrolyte solution undergo reduction and get discharged as metal atoms on the cathode surface. Simultaneously oxidation reaction should occur at the anode. Some of the oxidizable species present in the electrolyte solution may undergo oxidation at the anode. In such cases, electrolyte is continuously consumed in the electroplating process and is known as electrolyte consumption method of electroplating. Electrolyte need to be replenished periodically to maintain constant electrolyte concentration during electroplating. But, in certain cases the anode can be made of the same metal as it is deposited. For example, in the electroplating of copper, copper can be used as anode. In such cases, the anodic metal itself undergoes oxidation at the anode and dissolves in the solution and maintains a constant metal content in the electrolyte. This is known as anode consumption method of electroplating.

Electroplating being an electrolysis process, the amount of metal getting deposited and the amount of current passing through the electrolytic cell are related to each other by the laws of electrolysis, namely **Faraday's Laws**.

Faraday's first law of electrolysis: "The amount of substance deposited, dissolved or liberated at an electrode is directly proportional to the quantity of electricity passing through the electrolyte solution during electrolysis".

The quantity of electricity (Q) is equal to the product of current strength and the time for which it is passed. The unit of electricity is coulomb, which is the quantity of electricity passed through a conductor when a current of 1 ampere flows for one second. Therefore, the quantity of electricity in coulomb is the current in amperes multiplied by time in seconds.

$$\text{Coulomb} = \text{Ampere} \times \text{second}.$$

If W is the weight of a substance liberated or deposited at an electrode during electrolysis, then from the above law,

$$W \,\alpha\, Q$$

But Q = It, where I is current in amperes and t is time in seconds.

Therefore, $W \,\alpha\, It$ or $W = ZIt$, where Z is a constant of proportionality, known as electrochemical equivalent of the substance. When Q = 1 coulomb, W = Z. Therefore, electrochemical equivalent may be defined as the weight of the substance deposited/ evolved by the passage of 1 coulomb of electricity.

Faraday's second law of electrolysis: "When same quantity of electricity passes through solutions of different electrolytes, the amounts of substances liberated/deposited at the electrodes are directly proportional to their chemical equivalents."

$W \,\alpha\, E$, where W is the weight of the substance liberated/deposited and E is the chemical equivalent weight of the substance liberated/deposited.

If W_1 and W_2 are the weights of two different substances deposited by the passage of same quantity of electricity and $E_1$and E_2 are their respective chemical equivalent weights, then

$$\frac{W_1}{E_1} = \frac{W_2}{E_3}$$

The quantity of electricity which must be passed in order to deposit/liberate 1 gram equivalent weight of any substance is constant and this is called Faraday. 1 Faraday is equal to 96,500 coulombs.

Since 1 coulomb deposits or liberates 1 electrochemical equivalent and 96,500 coulombs deposit/ liberate 1 gram equivalent of a substance,

Chemical equivalent (gram equivalent) = 96,500 × electrochemical equivalent.

If the same amount of electricity is passed through two electrolytic cells containing the solutions of two different electrolytes AB and CD; and if the weights of A and C deposited/liberated on their respective electrodes are x and y grams, then

$$\frac{\text{Wt of A deposited}}{\text{Wt of C deposited}} = \frac{\text{Chemical equivalent of A}}{\text{Chemical equivalent of C}} \quad \frac{x}{y} = \frac{E_A}{E_C}$$

Thus knowing the weights of elements A and C deposited and equivalent weight of any one, the equivalent weight of the other can be calculated.

Example 4.1. Calculate the weight of silver deposited when a current of 2 ampere strength is passed for 20 minutes in an electrolytic cell containing silver nitrate solution. Chemical equivalent of silver is 108.

Solution:

Quantity of electricity passed = Ampere × time in s

= 2 × 20 × 60 = 2,400 coulombs.

96,500 coulombs deposit = 1 g equivalent weight of Ag

2,400 coulombs deposit $\frac{2,400}{96,500}$ g eq. of Ag

$\frac{2,400 \times 108}{96,500}$ g of Ag

2.686 g of Ag

Example 4.2. How many grams of oxygen can be produced by electrolysis of water using a current of 20 amperes strength for 10 minutes?

Solution:

Quantity of electricity passed = It = 20 × 10 × 60 = 12,000 coulombs.

96,500 coulombs liberate = 1 g equivalent of oxygen

= 8 g of oxygen

Therefore, 12,000 coulombs $\frac{8 \times 12,000}{96,500}$

= **0.995 g of oxygen.**

4.6. ELECTROPLATING PROCESS

The electroplating device is essentially an electrolytic cell, in which two electrodes, anode and cathode are dipped in an electrolyte solution. The principal components of an electroplating process are shown schematically in Fig. 4.3.

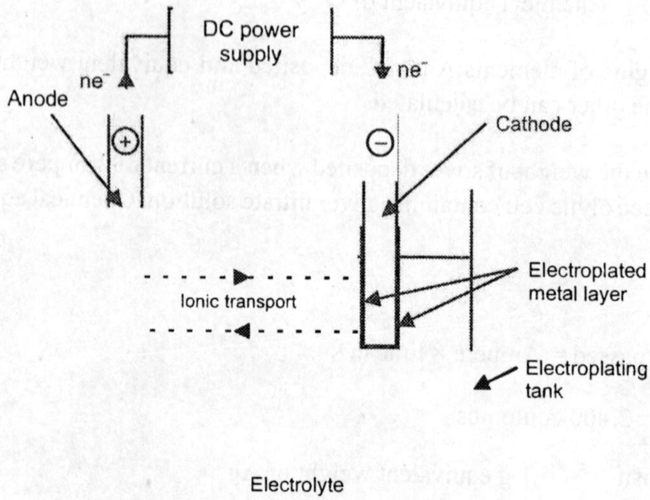

Fig. 4.3. Schematic representation of electroplating unit

The essential components include:

1. An electroplating bath containing a conducting salt and the metal to be plated in a soluble form as well as a buffer and additives.
2. The electronically conducting cathode, i.e., the article to be plated.
3. The electronically conducting anode, the coating metal itself or an inert material of good electrical conductivity like graphite.
4. An inert vessel to contain the above mentioned materials, made up of either rubber lined steel, plastic, concrete or wood.

Electrolysis reactions

Electroplating is the process of electrolytically depositing a layer of metal onto a surface. The object to be plated is made the cathode in an electrolytic bath containing a metal ion, M^{n+} so that the simplest reaction at the cathode is;

$$M^{n+} + ne^- \longrightarrow M$$

Wherever possible, the preferred anode reaction is the dissolution of the same metal in the solution.

$$M \rightarrow M^{n+} + ne^-$$

In such cases, there is a continuous replenishment of electrolyte during electrolysis.

However, if the anode is made of an inert material that does not pass into the solution, the electrolytic salt is added continuously in order to maintain optimum metal ion concentration in the solution.

As an example, in the electroplating of copper, the coating metal copper itself is made anode, which dissolves as Cu^{2+} ions, replenishing the Cu^{2+} ions deposited at the cathode. $CuSO_4$ solution can be used as electrolyte, which ionizes as

$$CuSO_4 \rightarrow Cu^{2+} + SO_4^{2-}$$

On passing current, Cu^{2+} ions migrate to the cathode and get deposited there.

At the cathode, $$Cu^{2+} + 2e^- \rightarrow Cu$$

At the anode copper enters into the solution as Cu^{2+} ions.

At the anode, $$Cu \rightarrow Cu^{2+} + 2e^-$$

Thus in this case, there is a continuous replenishment of electrolyte during electrolysis.

4.7. CHARACTERISTICS OF A GOOD DEPOSIT

The following are the characteristics of a good deposit:

- The deposit should be continuous, uniform, non porous and adhesive.
- It should be of fine grained nature.
- It should be bright and lustrous.
- It should be hard and ductile.

Electrodeposited metal consists of crystals, each of which, in turn, is made up of large numbers of lattice units. The properties of metal deposits are determined principally by the size and arrangement of the individual crystals. The type of crystals produced under given conditions, depends largely upon the relative rates of: (a) formation of nuclei from which new crystals are developed and (b) growth of existing crystals. The general conditions that favour: (a) will yield fine grain deposits, which contain more but smaller crystals while, conditions that favour (b) will result in fewer but larger crystals.

In general, for a given metal, fine grained deposits are smoother, brighter, harder, stronger but less ductile than coarse grained deposits. However, some exceptions do exist.

4.8. FACTORS INFLUENCING THE NATURE OF DEPOSIT

There are several factors which affect the nature of an electrodeposit. They are explained in the following sections.

(i) Current Density: Current density is the current per unit area of the electrode surface. It is expressed in milli amperes per cm^2 or amperes per dm^2 or amperes per m^2.

In general, it is desirable to use as high current density as practicable and thereby to increase the rate of electroplating. When current density is low, the metal is deposited slowly, providing adequate time for the metal nuclei to grow and with little scope for the creation of fresh nuclei. Thus the deposit produced is gross grained. Within certain limits, an increase in current density increases the rate of electrodeposition, decreases the crystal size and the deposit produced in that range will be fine grained and uniform. When the current density exceeds the limiting value for that bath and temperature, there is a tendency to produce rough and treed deposits due to the rapid discharge of ions on the cathode with a tendency to grow outward toward the bulk of the solution. When the current density is above the optimum limiting value, one more possibility is the formation of a burnt and spongy deposit. This is caused by the occurrence of hydrogen discharge at the cathode, increasing the hydroxyl content in the electrolyte, in the immediate vicinity of the cathode. As a consequence, metal hydroxides are precipitated. These, when included in the electrodeposit in adequate amounts, make the deposit to appear burnt, spongy and dark.

Therefore, an optimum current density should be applied to have a better deposit from a given bath.

(ii) Metal salt and electrolyte concentration: Normally, the metal salt concentration is kept high, because, a high current density can be employed at high metal content in the solution. In such a bath, the limiting current density is raised; the bad effects of high current density are thus shifted to higher values of current density. Various electrolytes are also added in high concentrations to increase the conductivity of the solution and cathode efficiency. Firm and adherent deposits can be obtained with higher electrolyte concentration and intermediate current density.

(iii) Metal ion concentration: As the metal deposition involves discharge of the free metal ions, the concentration of the metal ions is more important than that of the metal compounds. In general, a decrease in metal ion concentration decreases the crystal size and results in a fine adherent coating films. The low free metal ion concentration in a strong solution of metal compounds can be achieved either by the addition of a compound with a common ion or by the formation of complex compounds and ions.

For example, when copper is deposited from copper sulfate bath, sulfuric acid is added to the solution.

$$CuSO_4 \rightleftharpoons Cu^{2+} + SO_4^{2-}$$

$$H_2SO_4 \rightarrow 2H^+ + SO_4^{2-}$$

Due to the common ion effect of SO_4^{2-} from sulfuric acid, the concentration of cupric ions in the solution is reduced. The effective metal ion concentration can be reduced to about 10% of that present in that concentration of the metal salt alone.

The most important complex compounds used in plating are the double cyanides of sodium or potassium with metals such as copper, silver, gold, zinc and cadmium. In copper plating, a very low concentration of copper ions in the electrolytic bath is obtained by using a solution of sodium cuprocyanide, $[NaCu(CN)_2]$, obtained by dissolving cuprous cyanide in NaCN.

$$CuCN + NaCN \rightarrow Na[Cu(CN)_2]$$

$$Na[Cu(CN)_2] \rightleftharpoons Na^+ + [Cu(CN)_2]^-$$

$$[Cu(CN)_2]^- \rightleftharpoons Cu^+ + 2CN^-$$

Low concentration of cuprous ions is formed through the formation of $[Cu(CN)_2]^-$. This serves as a reserve of cuprous ions and releases cuprous ions continuously for the deposition of copper.

(iv) Agitation: In general, agitation of the solution brings up a fresh supply of metal salts or ions to the cathode and thus facilitates replenishment of the metal ions at the cathode surface. This provides higher current density and better current distribution and improves the deposit. Another advantage of agitation is that it sweeps away gas bubbles, which may otherwise cause pits.

The disadvantages of rapid agitation are that it may detach particles from the anode and stir up sludge, resulting in rough and porous deposits.

(v) Temperature: In general, an increase in the temperature of deposition causes an increase in crystal size. However a good, fine grained and smooth deposit is obtained at slightly higher temperatures, because of the following reasons. At higher temperature, the solubility and dissociation of the metal salt increases, this in turn, leads to a higher conductivity of the solution. A high temperature increases the mobility of the metal ions and decreases the viscosity of the solution so that the cathode film is more rapidly replenished. This reduces the tendency toward treeing and also increases the current density obtained with a given voltage. But the disadvantages of maintaining the higher temperature are the possibility of corrosion of process equipment, hydrogen evolution at the cathode and decomposition of organic addition agents.

(vi) pH of the electrolytic bath: For a good electrodeposit, the pH of the bath must be properly maintained at a definite range. In general, the cyanide baths are alkaline with pH values varying from about 9 for silver to about 13 for cadmium. For non cyanide baths, slightly basic or acidic conditions are preferred. At a higher pH, precipitation of hydroxides of the metal may take place. At very low pH, more hydrogen evolution takes place on the cathode, affecting the nature of the deposit. The required pH is maintained by using suitable buffers. For example, in nickel plating, borate buffer is used to maintain the pH between 4.0 and 4.5.

(vii) Addition agents: The plating bath normally contains certain electrolytes and various types of additives in addition to the salt of the metal to be plated, in order to obtain electroplate with desired properties and qualities. The wide ranges of additives added in low concentrations to the electroplating bath, modify the structure, morphology and properties of the electro deposit. The different types of additives used are discussed below.

(a) Complexing agents: Complexing agents are used for the following purposes:

- to maintain low metal ion concentration in the solution when the metal content of the solution is high.
- to prevent the chemical reaction between cathode metal and plating ions (e.g. in plating copper on iron, the reaction of Cu^{2+} ion with iron is prevented by complexing Cu^{2+}).
- to prevent the passivation of anode and consequent loss of current efficiency (e.g. low concentration of chloride ions is a common constituent of many baths).
- to improve the throwing power of the plating bath.
- to enhance the solubility of the slightly soluble metal salts.

The most common complexing agents used in electroplating are cyanide, hydroxide and sulfamate ions.

(b) Brighteners: In order to obtain a bright and lustrous deposit, specific chemicals are added to plating baths. These addition agents, called "brighteners" are employed in Ni, Zn, Cd, Sn, and other types of baths, which in their absence yield dull deposits. For example compounds like aromatic sulphones, sulphonates, thiourea, coumarin, etc., are used as brighteners in nickel plating.

The molecules of the brightener are adsorbed on the metal nuclei, preventing the freshly discharged ions from depositing over the earlier nuclei. This initiates the formation of new nuclei, i.e., more number of smaller crystals resulting in the formation of fine grained deposit. Certain organic compounds such as aryl sulfonic acid may not themselves produce bright deposits but they increase the brightening power of other compounds.

The concentration of the brighteners added should be maintained at an optimum range because excess amounts of brightener gives a fine grained deposit but brittle as large amount of the addition agent gets included in the deposit.

(c) Levelers: Certain organic compounds are used as levelers in order to get a level deposit. They produce a level deposit by preferentially getting adsorbed at places where rapid deposition of metal takes place. The adsorbed levelers will reduce the rate of deposition at those points. In practice, many brighteners also act as levelers.

(d) Structure modifiers: These additives change the structure of the deposit and the orientation or type of lattice. They are used to optimize particular deposit properties. Stress relievers are used to avoid the development of internal stress in the deposit.

(e) Wetting agents: Wetting agents are used to reduce or prevent pitting by reducing the tendency of hydrogen to stick to the cathode surface. Removal of hydrogen also avoids the possibility of hydrogen embrittlement of the deposit. Wetting agents, sometimes improve the uniformity of the deposit. Sodium lauryl sulfate is used as wetting agent.

In general, additives may affect more than one property of the deposit and also that when several additives are present in the electrolyte, their effect is synergistic.

(viii) Throwing power of the plating bath: The property, throwing power of a plating bath has been used as a measure of the uniformity in the distribution of the deposit on different parts of an article. It is defined as "the degree of uniformity of metal distribution or evenness of deposit thickness obtained on a cathode of irregular shape".

The property is particularly important when the article is of an irregular shape. If the distribution of the deposit is uniform through out the surface of the article to be coated, the throwing power is said to be good.

The throwing power of an electroplating bath is decided by the following factors:

(a) Conductivity of the electrolyte: In an electrolyte of high conductivity, the current distribution on the cathode surface is uniform and hence the metal is also deposited uniformly. Thus the solutions with higher conductivity will have higher throwing power.

(b) Presence of addition agents: The presence of additives like levelers and brighteners ensure an even deposit and hence increase the throwing power.

(c) Competing electrode reactions: Occurrence of competing reactions or side reactions at the cathode like hydrogen evolution increases the throwing power. Hydrogen evolution occurs at points on the surface where the potential or current distribution is high. The hydrogen evolution reduces the potential at those points, leading to a more even deposit.

Although throwing power is essentially a property of the solution, the following process parameters also affect it. They are, bath composition, pH, temperature, current density, geometry of the electrodes and their position relative to each other and the tank walls, type of cathode, etc.

Throwing power of plating bath can be determined by using the Haring-Blum cell (Fig. 4.4).

The cell consists of a box of PVC or similar insulating materials of 15 × 5cm dimension. Sheet metal cathodes are fitted at both ends of the box and are connected by a rod. An anode is placed between the cathodes at different distances x_1 and x_2 from the cathodes.

The cathodes are weighed initially. Electrodeposition is carried out for a known period of time. At the end of the test, the cathodes are weighed again and from this, the throwing power is calculated by using the equation.

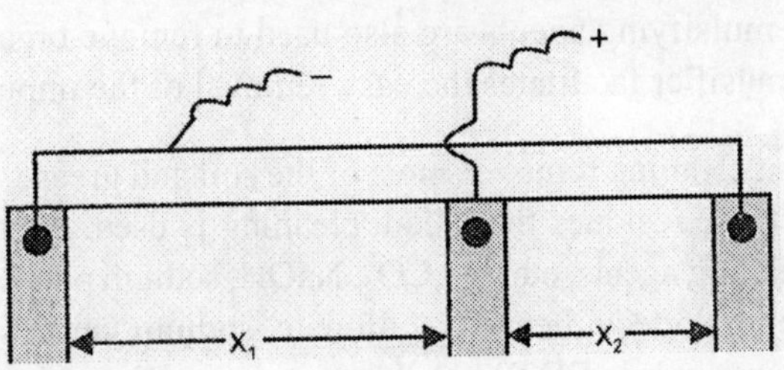

Fig. 4.4. Haring-Blum cell

$$TP\% = \frac{K - M}{K + M - 2} \times 100$$

where $K = x_1 / x_2$ and $M = w_2 / w_1$ where w_1 and w_2 are the weights of the metal deposited on the cathodes.

(ix) Cleaning of the articles to be plated: A good, adherent deposit is obtained only if the base metal surface is free from any dirt and extraneous matter such as grease and corrosion product. Any non metallic or poorly conducting particles on the surface are likely to prevent or retard the deposition at those points and hence to cause pores in the deposit. Therefore it is necessary to clean the article well, prior to electrodeposition.

Practically, in all cases, a given deposit produced on a rough surface will be less smooth than that on a fine surface. Hence, if a bright deposit is desired, it is necessary to use a metal surface with a fine finish.

4.9. METHODS OF CLEANING THE METAL SURFACE TO BE COATED

Prior to the application of electroplated coatings, it is essential to clean the surface of base metal. The common impurities found on the metal surface are grease, oil, oxide layer and other extraneous matters. The following methods are employed to clean the metal surfaces.

(i) Solvent cleaning: Solvent cleaning involves the cleaning of the metal surface by using organic solvents such as carbon tetrachloride, toluene, xylene, trichloroethylene, perchloroethylene, etc. These solvents remove the organic impurities such as oil, grease, etc.

A more effective way of solvent cleaning is by vapor degreasing. In this method the solvent like trichloroethylene is vaporized by heating and the vapors are made to condense on the metal surface to be cleaned. The condensed liquid dissolves and washes away the oil, grease and other organic matter from the surface.

Organic solvents with emulsifying agents are also used to remove organic impurities from the surface of the metal. The emulsifier facilitates the easy removal of the impurity from the surface.

(ii) Alkali Cleaning: Solvent cleaning removes most of the soil and grease, but minute residues are left behind. It is to remove these residues that alkali cleaning is used. Alkali cleaners consist of a number of alkalis, soaps, chelating agents, etc. Na_2CO_3, NaOH, sodium phosphates are the commonly used alkaline cleaners. Sodium dodecyl benzene sulfonate, sodium lauryl sulfate, etc., are used as wetting and emulsifying agents, while EDTA, sodium citrate, sodium gluconate, etc., are used as chelating agents.

Alkali cleaning is made more effective by passing current through a hot alkaline solution, with the article to be cleaned constituting the cathode or anode. The evolution of hydrogen at the cathode metal or oxygen at the anode metal, results in strong agitation, which helps to dislodge the greasy substances. Alkali cleaning is followed by thorough rinsing with water and then immersing in a very dilute solution of acid to remove traces of alkali.

(iii) Mechanical Cleaning: Mechanical cleaning involves removal of the oxide layer or rust and other inorganic deposits on the metal surface. The simple methods of mechanical cleaning involve the hand cleaning with bristle brush, and some abrasives like polishing tools, sand papers, etc. Cleaning may also involve grinding using polishing machines, which consist of polishing wheels, with a coating of abrasives like emery, carborundum, silica, etc.

Sand blasting method is often used when large surface needs to be cleaned. The process involves introducing the sand into an air stream under a pressure of 25–100 atmospheres and the blast is impacted on the metal surface to be cleaned. The method requires expensive equipments and involves health hazards to workers, because of the possibility of *silicosis*, unless special protective measures are taken.

(iv) Pickling: The methods used for the removal of oxides generally involve their dissolution by means of an acid. The type and concentration of acid, its temperature and the time of treatment vary with the kind and the amount of oxide to be removed. In general, the term pickling refers to severe or continued treatment, while the term dipping refers to a short treatment.

Pickling of steel involves the dipping in dil HCl or dil H_2SO_4 solution which contains some inhibitors which retard the dissolution of steel itself. For copper and its alloys, the pickling solution is dil H_2SO_4 or a mixture of dil HNO_3 and dil H_2SO_4. Aluminium and its alloys are descaled by alkali pickling.

(v) Electropolishing: Electropolishing involves anodically pickling the metals in order to have a polished metal surface for electroplating. In this method, the metal to be cleaned is made as anode in a suitable solution. During the process, a surface layer of the metal gets dissolved along with the impurities. The method also removes minor surface irregularities. The most commonly used baths for electropolishing contain sulfuric acid, phosphoric acid, chromic acid, nitric acid, etc.

After the surface cleaning processes, the metal is thoroughly rinsed with water, dried and used for electroplating purpose.

4.10. REQUIREMENTS OF AN ELECTROLYTE SOLUTION FOR ELECTROPLATING

The electrolyte solution used for electroplating should possess the following characteristics:

- Metal content of the solution should be high and the free metal ion concentration should be relatively low.
- The conducting power of the solution should be high.
- It should be stable under the operating conditions, without undergoing hydrolysis, oxidation, reduction and other chemical changes.
- The solution should dissolve the anode quite efficiently whenever an active anode is used and should be able to maintain constant metal content of the solution.
- It should possess good covering power and throwing power. Covering power is the measure of the ability of the bath solution to deposit the metal over the entire surface of an article.

However, no single electrolyte satisfies all these requirements. Therefore, usually, a mixture of two or more electrolytes is used along with suitable addition agents.

4.11. APPLICATIONS OF ELECTROPLATING

(i) For better appearance (Decorative plating) : To increase the commercial and decorative values, base metals are electroplated with gold, nickel, chromium, silver, platinum, palladium, etc.

(ii) Plating for protection: Electroplating is widely used to have protective coating over a metal or alloy which is susceptible for corrosion. For example, iron and steel are protected from corrosion by electroplating with zinc, nickel, cadmium, etc. Where both protection and good appearance are desired, more than one metal are plated on a metal surface. For example, steel is given successive coating layers of copper, nickel and chromium. Copper and nickel provide protection against chemical attack and rusting, while chromium gives a better appearance.

(iii) Plating for special surfaces: A part of a plant can be fabricated with an alloy which gives all desirable properties, but is susceptible for corrosion or wear under the working conditions. In such a case, the part requiring corrosion resistance or wear resistance is exclusively electroplated with a suitable metal. For example, in internal combustion engines, wear can be avoided and running performance can be improved by applying an electroplated chromium layer.

(iv) Plating for engineering effect: Electroplating is employed for temporary use in metal treatment. For example, steel parts are copper plated before carburizing to prevent carburization at undesired portions. In hardening process, steel is electroplated with tin or copper-tin alloys at regions where nitriding is not required.

(v) For electroforming : Electroforming is a method of forming or fabricating an object by electroplating for an extended period, for specialized applications. Examples of electroforming are wave guides and other structures used in electronics, reflectors, pen caps, musical instruments, copper floats, missile parts, moulds and dies for rubber, plastic, etc., paint spray masks, phonograph record masters and stampers, electrotypes and other items used in the printing industry, foils and sheets for different applications, filter screen meshes, etc.

(vi) Plating on non-metallics : Non-metallics are electroplated with metals for decorative look and to impart functional properties. Plated plastics are found in automobiles, radio and television sets, furniture and domestic items, electrical and textile industries, marine hardware, plumbing fixtures and accessories, consumer items such as ball-point pens and clocks.

4.12. ELECTROPLATING OF CHROMIUM

Chromium is also one of the widely used metals for electroplating. In the field of electroplating two types of chromium are employed: decorative and hard.

(i) Decorative Chromium: A thin deposit of chromium (0.25–0.75 μm) is applied over either copper-nickel or nickel undercoat. It confers a pleasing look and non-tarnishing nature to the plated article.

Bath Composition :	250 g of chromic acid + 2.5g of H_2SO_4 (Ratio of chromic acid to H_2SO_4 is 100:1) pre liter of the solution. Should contain 1g of trivalent chromium, produced in situ by adding 4g of oxalic acid.
Operating temperature	45–55°C
Current density	145–430 A/ft^2
Current efficiency	10–15%
Anode	Insoluble anodes like lead, lead – 6% antimony, lead –7% tin are used.
Cathode :	Article to be plated.

Mode of plating: Generally a thin coating of chromium is applied which contains considerable number of pores. As the thickness of decorative coating increases, the coating tends to develop cracks. Therefore chromium coatings are plated on nickel under coat, so that chromium gives decorative finish and nickel gives protection. For ferrous articles an undercoat of copper is also employed below nickel.

The reasons for this are as follows:

- The nature of finish obtainable with copper is better than on steel.
- The coverage of nickel and hence of chromium is better with copper below.
- Specifications permit a lower thickness of nickel if a specified copper undercoat is given.

(ii) Hard chromium: Hard chromium plating (Industrial or engineering chromium) involves deposition of a thick coating of chromium directly over the substrate. The usual thickness range of hard chrome is 2.5–300 μm.

Bath composition :	250g of chromic acid + 2.5g of H_2SO_4 and 1g of trivalent chromium per litre of the bath.
Operating Temperature :	45–55°C
Current density :	290–580 A/ft^2
Current efficiency :	17–21%
Anode :	Lead, lead – 6% antimony, lead – 7% tin anodes
Cathode :	The article to be coated
Hardness :	835 – 925 VPN

In the case of chromium plating inert anodes are used instead of soluble chromium anodes. This is because of the wide difference between the anode and cathode efficiencies. The anode efficiency is nearly 100%, but cathode efficiency is at best around 20%. This means the metal going into the solution is five times of the metal getting deposited, resulting in building up of excessive chromic acid concentration. This leads to imbalance of bath composition with respect to Cr(III) – Cr(VI) ratio and chromic acid-sulfuric acid ratio, both of which are important factors in deciding the nature of the deposit. Also, there is a change in bath conductivity. As a result of these changes, a poor quality chromium deposit is obtained.

Chromium plating produces an attractive, wear resistant and abrasion resistant surface. Hard chrome deposit confers to the plated surface a combination of physical and mechanical properties, such as high hardness, abrasion resistance, low coefficient of friction, good-corrosion resistance, high heat resistance (up to 400°C), non sticking and antiseizing properties.

Applications

- Decorative chromium provides a durable finish on cycles, automobiles, furnitures, household fittings, aircraft, surgical and dental instruments.
- Hard chromium is applied to many components in almost every industry. Some of them are gauges, dies, cutting tools, piston rings, cylinder liners, crankshafts of marine and aero engines, bearings, hydraulic rams, and in printing industry.
- Hard chrome is also used in building up of worn-out or mismachined parts.
- Porous chromium plating is employed on piston rings and cylinder liners of diesel engines and aircraft engine cylinders.
- Black chromium is used in optical instruments, machine tools, and electronic parts.
- It is also used for non-glare finishes on automobiles, and as an efficient coating for solar energy collectors.

4.13. ELECTROPLATING OF GOLD

Gold plating was initially applied for decorative purposes on jewelry articles. But latter it found applications in the industries, for various purposes.

Typically four bath compositions are used for gold plating. Two of them are alkaline cyanide baths, one is neutral cyanide bath and the fourth one is acidic cyanide bath.

Rubber or plastic lined tanks can be employed for all of the baths, the temperature being kept below 70°C. Stainless steel tanks may also be used for alkaline cyanide baths. Agitation in terms of cathode movement or solution circulation is desirable. The details about the bath compositions and conditions of plating are given in Table 4.2.

Bath A and B are used primarily for decorative gold plating and for industrial uses except for electrical contact requiring wear resistance. Bath A is used for low thickness of gold (0.02 to 0.25 μm) and bath B is used for large thickness of gold (up to 40 μm for items subject to heavy wear and abrasion). These baths are not suitable for plating on printed circuit boards.

Bath C is used for components where baths A and B can not be used. The gold deposits from this bath are less porous compared to those from baths A and B. Bath D is similar to bath C in use and yields a gold deposit of 99.999% plus purity with highest ductility.

Since insoluble anodes are used in almost all gold baths, gold salt – the potassium gold cyanide complex, is periodically added to replenish the gold.

Table 4.2

Constituents	*Bath A (alkaline)*	*Bath B (alkaline)*	*Bath C (neutral)*	*Bath D (acidic)*
Potassium gold cyanide (g dm^{-3})	1.5–3.0	8.0	6.0	6.0–18.0
Potassium cyanide (g dm^{-3})	7.5	20.0		
Potassium carbonate (g dm^{-3})	20.0	20.0		
Dipotassium monohydrogen phosphate (g dm^{-3})	15.0	20.0	20.0	
Monopotassium dihydrogen phosphate (g dm^{-3})			15.0	20.0
Potassium citrate				50.0
pH	11–13	11–11.5	6.5–7.5	3–6
Temperature (°C)	60–70	50–60	25–70	40–70
Current density (A / ft^2)	1–5	1–5	5	1–20
Cathode efficiency (%)	100	100	95–100	80–90
Anode	Pt, Stainless steel, gold	Pt, Stainless steel, gold	Pt, Platinised Ti, Stainless steel	Pt, Platinised Ti, Carbon

When gold is directly plated on copper, the copper atoms have the tendency to diffuse through the gold layer, causing tarnishing of its surface and formation of an oxide layer. Therefore, a layer of a suitable barrier metal, usually nickel to be deposited on the copper substrate, before gold plating. The nickel deposit also provides mechanical backing for the gold layer, improving its wear resistance. It also reduces the effect of pores present on the gold layer. When gold is plated on silver, silver atoms also diffuse through the gold layer, causing gradual fading of its colour and tarnishing the surface. This process may take months to years, depending on the thickness of the gold layer. Therefore, it is better to use a barrier metal layer underneath the gold deposit.

Applications

Gold deposit is a lustrous deposit with good tarnish and corrosion resistance. It is with good electrical conductivity, comparable with copper and silver, and has an ability to reflect IR radiation.

- Used for decorative purposes in jewelry, watch cases, pen points, hollow ware, etc.
- In the electrical industry, printed circuits, contacts and connectors are gold plated.
- In electronics, transistors and integrated circuit parts are gold plated.
- In the aerospace industry, instrument coverings and external surfaces exposed to radiation in space are gold plated.
- Reactors and heat exchangers in the conductor tube for desalination are gold plated, because of the high corrosion resistance nature of gold.

4.14 ELECTROLESS PLATING

Electroless plating is a method of depositing a metal or alloy over a substrate (conductor or non-conductor) by controlled chemical reduction of the metal ions by a suitable reducing agent without using electrical energy.

The reduction of metal ions by the reducing agent is catalyzed by the metal atoms being plated. Therefore, electroless plating is also termed as autocatalytic plating.

The surface to be plated should be catalytically active so that in the beginning of the plating reaction, the surface of the substrate catalyses the reduction. Once the substrate is covered by the metal or alloy coating, the coating catalyses the reduction. The process continues with building of successive layers of the metal or alloy.

The electroless plating process can be represented as,

$$\text{Metal ions} + \text{Reducing agent} \xrightarrow{\text{Catalyst surface}} \text{Metal} + \text{Oxidized product}$$

The catalytic metals such as Ni, Co, Steel, Fe, Rh, Pd, Al, etc., do not require any surface preparation before electroless plating on them. Non-catalytic metals such as Cu, brass, Ag, etc., need activation. This can be achieved by rendering them cathodic for a short period by passing direct current or by dipping in palladium chloride – hydrochloric acid solution. Non-conductors like glass, plastics, ceramics, etc., are first activated in a solution of $SnCl_2$ and HCl. After rinsing, it is immersed in a solution of $PdCl_2$ and HCl.

Comparison of electroplating and electroless plating

Sl.No.	*Property*	*Electroplating*	*Electroless plating*
	Driving force	Power supply	Autocatalytic redox reaction
2	Cathode reaction	$M^{n+} + ne^- \rightarrow M$	$M^{n+} + ne^- \rightarrow M$
3	Anode reaction	$M \rightarrow M^{n+} + ne^-$ Or $n/2\ H_2O \rightarrow n/4\ O_2 + nH^+ + ne^-$	$R \rightarrow Ox + ne^-$
4	Site of cathode reaction	Article to be plated	Article to be plated which must have a catalytic surface
5	Site of anode reaction	Separate anode	Article to be plated
6	Anode reactant	M or H_2O	R, reducing agent in solution
7	Nature of deposit	Pure metal or definite alloy	Usually metal contaminated with O/R derived species
8	Thickness limit (μm)	1–100	1–100

Composition of electroless plating baths

An electroless plating bath in general will contain the following components:

(i) Metal salts to provide metal ions for deposition.
(ii) Reducing agent for reduction of metal ions to metal atoms.
(iii) Complexing agents to complex metal ions to prevent bulk deposition.
(iv) Stabilizer to give more stability to the solution.
(v) Exaltent or accelerator to increase the rate of plating.
(vi) Buffer to control pH.
(vii) Other additional agents like brighteners.

* The selection and concentration of complexing agent must be considered very carefully; because if the metal is too-heavily complexed, sufficient free metal ions will not be available for deposition.

Advantages of electroless plating

The advantages of electroless plating over electroplating process are as follows:

- Does not require electrical power source and accessories.
- Electroless plating baths have better throwing power and deposit a more uniform metal coating over an article irrespective of its shape or size. The only requirement is that the solution has free access to the areas to be plated.
- Electroless plating is applicable to conductors, semiconductors and nonconductors like plastics.
- Electroless deposits are less porous than electroplates and in general possess unique characteristic chemical, mechanical and magnetic properties.

Among the various metals applied by the electroless technique, nickel and copper are important in view of the extent of their applications.

4.15. ELECTROLESS PLATING OF NICKEL

Two typical bath compositions, one in acidic medium and the other in alkaline medium with operating conditions are given below:

Acid Nickel : $NiCl_2$ (20g), Sodium hypophosphite (20g), Sodium acetate (l0g), sodium succinate (15g) per liter, pH (4.5), Temperature (93°C)

Alkaline Nickel : $NiCl_2$ (30g), Sodium hypophosphite (l0g), Ammonium citrate (65g), Ammonium chloride (50g) per litre. pH (8–10), Temperature (90°C).

Reactions:

Cathode : $Ni^{2+} + 2e^- \rightarrow Ni$

Anode : $H_2PO_2^- + H_2O \rightarrow H_2PO_3^- + 2H^+ + 2e^-$

Over all reaction : $Ni^{2+} + H_2PO_2^- + H_2O \rightarrow Ni + H_2PO_3^- + 2H^+$

The nickel deposit formed contains phosphorous to the extent of 3–15%. The article to be coated is cleaned properly before applying the electoless metal coating and activated if necessary. The pH of the solution changes during plating. Therefore, buffers have to be added to maintain the pH. These include carboxylic acids in acid media and ammonia + ammonium chloride or organic amines in alkaline solutions.

The acid as well as alkali baths vary widely in their formulation, the constituents may be combinations of the following:

(i) Nickel salt : $NiCl_2$ or $NiSO_4$
(ii) Reducing agent : Sodium borohydride, dimethylamino borane, hydrazine hydrate or formaldehyde
(iii) Complexing Agent : Organic acids like lactic acid, propionic acid or salts (Also as buffers) like trisodium citrate
(iv) Stabilizers : Organic and inorganic thio compounds, metal ions like lead, thallium, etc.. thiourea, mercaptobenzothiazole.
(v) Exaltants : Lactates, succinates, hydroxy acetates, glycine, fluorides, maleonates, etc.
(vi) Brighteners : Sodium benzene disulphonate, thiourea, lead ions, etc.

The electroless plating is costlier than electroplating due to the use of expensive chemicals. Electroless nickel plate exhibits following characteristic properties:

- Good adhesion
- High hardness
- Non porous and high corrosion resistance, superior to pure nickel plate
- Solderable and weldable
- Low coefficient of friction and good wear resistance
- Very low ductility

Applications

- Because of their hardness and abrasion resistance, used on industrial components such as pumps and valves, pistons and shafts, gears, parts of hydraulic systems, reaction vessels, tools and dies.
- Because of their protective value, in screw conveyors, extrusion presses, filters, heat exchangers, etc.
- Also used in instrumentation and computers in view of its magnetic properties.
- Employed to coat steel, metallise plastic parts and printed circuit boards.

Limitations

- High chemical cost
- Slow coating rate (2–25 mmh^{-1})
- Poor stability of the solution
- Careful analytical control of the bath is required.

4.16. ELECTROLESS PLATING OF COPPER

The surface to be plated is cleaned free of impurities and activated if required.

Bath compositions

Constituents	*Bath A*	*Bath B*
Copper sulfate (gdm^{-3})	12	30
Rochelle salt (gdm^{-3})	55	140
Sodium hydroxide (gdm^{-3})	15	40
Sodium carbonate (gdm^{-3})		25
EDTA–sodium salt (gdm^{-3})		17
Formaldehyde (37% solution) ($ml.dm^{-3}$)	35	170
Stabilizer: Thiourea (gdm^{-3})	0.1	
Methanol ($ml.dm^{-3}$)		125
pH	11.5–12.0	11.5–12.0
Temperature	Room temperature	Room temperature
Deposition rate (μm/h)	2.5	20

Bath A is used for metallising non conductors and bath B for a rapid deposition rate, but bath B has poor stability.

Control of metal content and pH are important. To raise the pH, 100gdm^{-3} sodium hydroxide solution is used.

Reactions

Cathode $Cu^{2+} + 2e^- \rightarrow Cu$

Anode $2HCHO + 4OH^- \rightarrow 2HCOO^- + 2H_2O + H_2 + 2e^-$

Overall reaction $Cu^{2+} + 2HCHO + 4OH^- \rightarrow Cu + 2HCOO^- + 2H_2O + H_2$

Applications

- Widely used for metalising printed circuit boards.
- For producing through-hole connections.

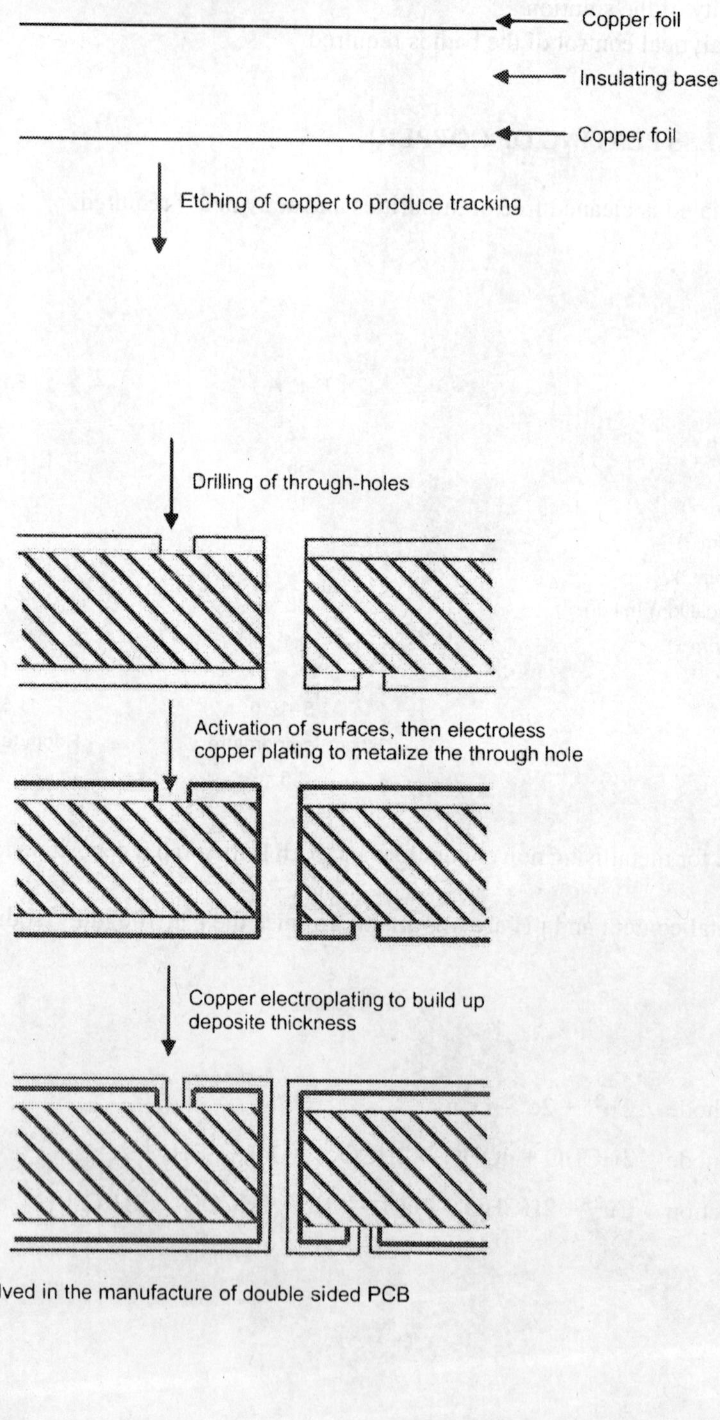

Fig. 4.5. Steps involved in the manufacture of double sided PCB

- For plating on non conductors.
- As a base for subsequent conventional electroplating.
- Applied on wave guides and for decorative plating on plastics.

Through hole connections are necessary when double sided printed circuit boards are fabricated. The electrical connections between two sides of the board are made by drilling hole and then plating through holes by electroless Plating.

For the manufacture of double sided printed circuit boards, the plastic board is initially covered with copper layers by cladding thin layers of electroformed copper foils. The copper clad board is printed with an etch- resistant pattern of the circuit or the tracking required on both the sides. On etching with suitable etchent the copper except below the printed pattern is etched away, leaving the circuit pattern. The connection between two sides is made by drilling hole followed by activation and plating through holes by electroless plating. The steps involved in the process is shown in Fig. 4.5.

REVIEW QUESTIONS

1. Enumerate the objectives of metal finishing.
2. Name the important techniques of metal finishing.
3. Define electroplating.
4. Explain the term polarisation. What are the types of polarisation?
5. What is decomposition potential? How is it determined? What is its significance in electrolysis?
6. What is discharge potential on an electrode? What is its significance in electrodeposition?
7. Define overvoltage. What are the factors influencing?
8. What is hydrogen over voltage? What is its importance in electroplating?
9. Discuss the principle of electroplating.
10. What are the characteristics of a good deposit?
11. Discuss, the different factors affecting the nature of the deposit
12. Explain the functions of following addition agents in electroplating:
    (i) Complexing agents
    (ii) Brighteners
    (iii) Levellers
    (iv) Wetting agents
13. What is throwing power of the plating bath? What are the factors affecting? How is it determined?
14. Explain the methods of cleaning a metal surface prior to plating.
15. What are the essential requirements of an electrolyte solution for electroplating?
16. What are the applications of electroplating?
17. Give brief accounts and applications of the electroplating of
    (i) Chromium
    (ii) Gold
18. What is electroless plating? How is it different from electroplating?
19. Explain the principle involved in electroless plating. What are the advantages?
20. Give an account of electroless plating of nickel and copper. What are the applications?
21. Account for the following:
    (i) Concentration polarisation can be eliminated by stirring the solution.
    (ii) Continuous electrolysis of water takes place only when the applied e.m.f is 1.7V and above.
    (iii) Zn and copper can be separated from a mixture of their solutions electrolytically.

(iv) Though the standard electrode potential of zinc is less than that of hydrogen, zinc can be electro deposited from acidic solutions.
(v) In the electro deposition of copper from CuCN, NaCN is added to the electrolyte bath.
(vi) During electroplating, the bath is maintained at slightly higher temperature.
(vii) Non cyanide baths are maintained at slightly acidic or slightly basic conditions.
(viii) Use of wetting agents in the bath solution avoids the formation of pitted deposits.
(ix) Before electroplating, the surface to be plated needs to be cleaned.
(x) To electroplate steel with copper from acid sulfate bath, cyanide copper strike is necessary on steel surface.
(xi) Decorative chromium is generally electroplated on nickel under coat.
(xii) In chromium electroplating of ferrous articles copper under coat is used below nickel undercoat.
(xiii) Chromium anode is not used in chromium electroplating.
(xiv) Electroless plating is known as autocatalytic plating.
(xv) In an electroless plating bath, presence of a reducing agent is a must.
(xvi) Copper is undercoated with nickel, before electroplating with gold.

22. State Faraday laws of electrolysis.

CHAPTER 5

Fuels (Energy Sources)

5.1. INTRODUCTION

One of the characteristics of a complex society is a high level of energy consumption. The per capita energy consumption is considered to be the measure of living standards of people. Just some 150 years ago, fuels were used primarily for cooking, heating and lighting purposes. Apart from complex cooking and heating, today fuels are used in transportation, communication, illumination, manufacturing and a host of other applications.

Coal, oil and natural gas, which are presently the principal sources of energy, are called fossil fuels (non-renewable energy sources). These fossil fuels are thought to have been formed over millions of years ago. Until the industrial revolution, fire wood was the main source of energy. As the wood sources depleted due to extensive and reckless deforestation, coal (black diamond) took the place of wood as a source of energy in industries and transportation. During the period of 1920–1930, the consumption of natural gas and petroleum increased rapidly due to the expanding use as fuel for internal combustion engines.

With an enormous rise in world population during the past few decades, the energy requirements have also increased at even larger rate. Our increasing appetite for energy is one of the causes for our energy crisis. According to some projections, we may run out of fossil fuels, by 2025. A considerable interest has been evinced in developing an alternate energy source. Much research has been focused on 'nuclear' and 'solar energy'.

Among alternate energy sources, nuclear fission is slowly replacing fossil fuels as a potential energy source of electric power. It generates enormous amount of energy but at the same time produces highly radio active materials that must be stored carefully in fail-proof containers for centuries. For these and other reasons, as the promise of cleaner, safer, more economical and practically inexhaustible energy source, though yet to be developed, controlled thermonuclear fusion can not be ignored.

5.2. CHEMICAL FUELS

Definition

A chemical fuel is a substance, which on combustion in air or oxygen produces significant amounts of heat, which can be conveniently used for useful domestic and other purposes. The principal elements invariably present in chemical fuels are carbon and hydrogen. When a chemical fuel undergoes combustion in air, carbon and hydrogen are converted into carbon dioxide and water respectively. Since the heat content of combustion products being lower than that of reactants, the chemical fuels release heat during their combustion process.

The examples of chemical fuels include wood, coal, crude oil, natural gas, etc. Chemical fuels are primarily used as a source of heat and power. Apart from this, they are also used as reducing agents in metallurgical applications.

Classification of Fuels

Chemical fuels are broadly classified as solids, liquids and gases based on their physical state in which they are used. These are further classified as primary and secondary fuels.

A fuel that occurs in nature is called a "**primary fuel**" while all others, which are derived from primary fuels are called **secondary fuels**. A complete classification of chemical fuels with suitable examples is summarized in Table 5.1.

Table 5.1. Classification of chemical fuels

Physical state	*Primary fuels (Natural)*	*Secondary fuels (Derived)*
1. Solid	Wood, Coal (lignite, bituminous, anthracite), Peat	Charcoal, coke
2. Liquid	Petroleum (Crude oil)	Gasoline, Diesel, Kerosene, Synthetic petrol
3. Gas	Natural gas	Producer gas, Water gas, Coal gas, LPG, Biogas

5.3. IMPORTANCE OF HYDROCARBON FUELS

Petroleum, coal and natural gas are excellent hydrocarbon fuels. They are called 'fossil fuels' (nonrenewable energy sources) formed from plant and animal remains buried deeply under high pressure, about 345 million years ago. These fuels are excellent hydrocarbon fuels as they contain carbon and hydrogen as major elements. When they are burnt in adequate supply of air, they are oxidized to CO_2 and water and release significant amount of energy in the form of heat which can be converted into a suitable form to perform useful work. Among fossil fuels, natural gas is relatively inexpensive. It is superior fuel and have many advantages over coal and petroleum. It burns cleanly, leaves no residue and produces less CO_2 per unit energy than the other two fossil fuels. Further, unlike coal and petroleum, it produces no NO_x, SO_x, CO, etc., as pollutants. It is therefore, the safest and excellent fuel available. The present day world is heavily dependent on coal, petroleum and natural gas for most of the energy needs such as transportation (vehicles, trains, aeroplanes), industrial processes, heating, cooling buildings (refrigeration) and generating electrical power (Thermal power plant).

Eighty percent of the world's energy demand is supplied by these hydrocarbon fossil fuels. The reserves of these fuels are fast depleting because of the high rate of consumption. One thing is certain: sooner or later the earth's store of fossil fuels will be used up and the sun is the ultimate inexhaustible source of energy.

5.4. CHARACTERISTICS OF A GOOD FUEL

Fuels are sources of energy. They are primarily used to produce heat energy that can be used for generating steam, power and heating. The performance of a fuel largely depends on its calorific value and also how clearly it burns. The following are the desirable properties of a good chemical fuel.

(i) High calorific value: The calorific value of a fuel is the direct measure of its efficiency as a fuel. Higher the calorific value, more efficient is the fuel. A good fuel therefore, must have a high calorific value.

(ii) Moderate ignition temperature: Ignition temperature is the minimum temperature to which the fuel is to be heated to start combustion. The fuels having very low ignition temperature can cause fire hazards during their handling, applications, storage and transportation. It is rather difficult to ignite a fuel if its ignition temperature is too high. Moderate ignition temperature is the most desired property of a fuel.

(iii) Low ash content: Formation of ash during combustion is due to the presence of inorganic matter in the fuel. High ash content, apart from lowering the calorific value, poses problems in its disposal. A good fuel must therefore, be free from ash content.

(iv) Low moisture content: The presence of high percentage of moisture in fuel decreases the efficiency of the fuel. It also increases the ignition temperature and fuel cost. Further, it lowers the calorific value since some of the heat produced is utilized to vaporize the moisture. Therefore the moisture content in a good fuel must be at negligible level.

(v) No harmful combustion products: It is one of the valuable properties of a good fuel. A good fuel must not produce harmful combustion products such as CO, SO_2, NO, H_2S, smoke and clinkers during combustion. Formation of such by-products can cause harmful effects on health. A good fuel must therefore burn with clean flame without producing undesired by-products.

(vi) Combustion control: One can avoid a large wastage of valuable fuel if its combustion rate can be properly regulated and burning can be stopped immediately as and when desired.

(vii) Ease of availability: Fuel must be readily available in abundant amount and its cost must be minimum.

(viii) Ease of storage and transport: Fuels should not demand more space and sophisticated arrangements for their storage and transportation. Further there must be minimum risks of fire hazards.

5.5. CALORIFIC VALUE

An important parameter of fuels is calorific value. It gives useful information about its heating efficiency. The performance of a fuel is expressed in terms of its calorific value. Calorific value of a fuel is defined as the amount of heat released when a unit quantity (mass or volume) of a fuel is burnt completely in air or oxygen.

The calorific value is normally expressed in calori per gram (cal/g) in cgs units. It is also expressed in joules per kg (J/kg) for solid fuels, and joules per cubic meter (J/m^3) for gaseous fuels in SI units.

Gross (Higher) calorific value

It is defined as the amount of heat released when unit quantity of a fuel is burnt completely in air and the combustion products are cooled to room temperature.

All fuels invariably contain carbon and hydrogen. Carbon and hydrogen present in fuels are converted into carbon dioxide and steam respectively, on combustion. On cooling the combustion products, steam gets condensed to water and liberates its latent heat. The measured gross calorific value includes the latent heat of steam. Therefore it is always higher than the net calorific value.

Net (lower) calorific value

It is defined as the amount of heat released when a unit quantity of a fuel is burnt completely in air, and the products of combustion are let off into the atmosphere. In actual practice, the combustion products are not cooled to room temperature, but simply let off into the atmosphere. Since this calorific value does not include the latent heat of steam, net calorific value is always lower than gross calorific value.

Net calorific value = Gross calorific value – latent heat of steam

$$= GCV - \frac{9 \times \%H}{100} \times 587$$

because 1 g hydrogen gives 9 g of water and the latent heat of steam is 587 $cal.g^{-1}$.

Bomb Calorimeter

Calorific value of solid or liquid fuel is determined by using a bomb calorimeter shown in Fig. 5.1.

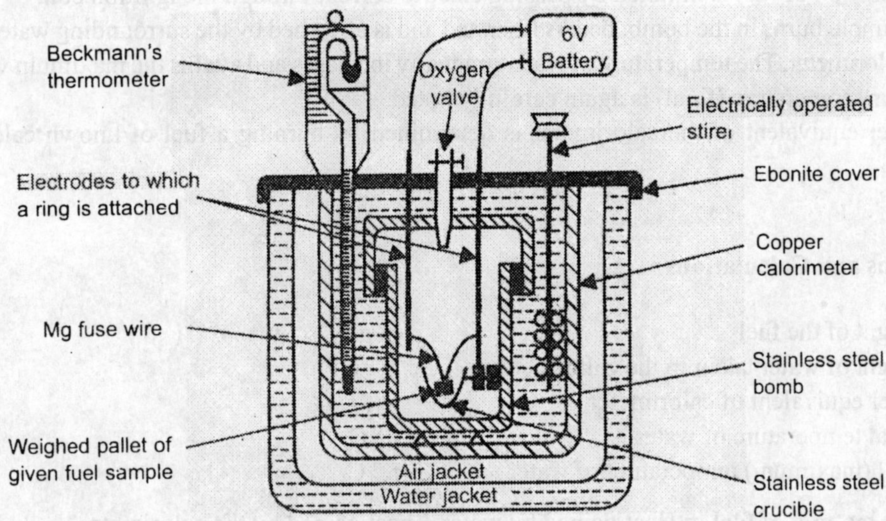

Fig. 5.1. Bomb calorimeter

Principle

A known weight of the sample (solid or liquid fuel) is burnt completely in excess of oxygen. The liberated heat is absorbed by the surrounding water and the calorimeter. Thus the heat liberated during the combustion of fuel is equal to the heat absorbed by water and copper calorimeter. The higher calorific value of the fuel is calculated from the data.

Construction

It consists of a stainless steel air tight sealed cylindrical bomb. The bomb has an inlet valve for providing oxygen atmosphere inside the bomb and an electrical ignition coil for the initiation of combustion of fuel. The bomb is placed in a large, well insulated copper calorimeter. The copper calorimeter is equipped with a mechanical stirrer for dissipation of heat and a thermometer to read accurately the temperature rise.

Working

A known weight of fuel (solid or liquid) is placed in a small stainless steel crucible. The crucible containing the fuel is next placed inside the bomb. The bomb is sealed air tight by the lid. The sealed bomb containing the sample is next placed in a large, well insulated copper calorimeter. It is covered by accurately measured quantity of water. The water is kept in constant agitation by the mechanical stirrer. The initial temperature of the water is carefully measured. The bomb is filled with oxygen and the combustion of fuel is initiated by passing electric current through the ignition coil.

As the sample burns in the bomb, heat is liberated and is absorbed by the surrounding water and the copper calorimeter. The temperature of water gradually increases and attains the maximum value. The maximum temperature (final) is again carefully noted.

The water equivalent of the calorimeter is determined by burning a fuel of known calorific value.

Observations and Calculations

(i) Weight of the fuel ... = m g
(ii) Weight of water taken in the calorimeter ... = W_1 g
(iii) Water equivalent of calorimeter ... = W_2 g
(iv) Initial temperature of water ... = t_1 °C
(v) Final (maximum) temperature of water ... = t_2 °C

Heat lost by 'm'g of fuel = Heat gained by water + heat gained by the calorimeter

$$= (W_1 + W_2)(t_2 - t_1) \text{ cal}$$

Higher calorific value of the fuel $= \dfrac{(W_1 + W_2)(t_2 - t_1)}{m}$ cal per gram

$$= \frac{(W_1 + W_2)(t_2 - t_1)}{m} \times \frac{4.184}{10^{-3}} \text{ J kg}^{-1}$$

Calculation of lower (net) calorific value

Let the fuel contain 'H' percentage of hydrogen.

$$2H + \tfrac{1}{2} O_2 \rightarrow H_2O$$

2g of hydrogen produces 18g of water.

Water formed by 1 g of fuel $= \dfrac{18}{2} \times \dfrac{H}{100} = 0.09\,H\,g$

Latent heat of steam = 587 cal / g

Latent heat of water formed = 0.09H × 587 cal g^{-1}

Lower (Net) Calorific Value = HCV – Latent heat of water formed

= HCV – 0.09H × 587 cal g^{-1}

Example 5.1

Calculate the gross and net calorific value of a coal sample from the following data obtained from bomb calorimeter experiment.

(i)	Weight of coal (m)	0.73 g
(ii)	Weight of water taken in calorimeter (w)	1500 g
(iii)	Water equivalent of calorimeter (w_1)	470 g
(iv)	Initial temperature (t_1)	25.0°C
(v)	Final temperature (t_2)	27.3°C
(vi)	Percentage of Hydrogen in coal sample	2.5 %
(vii)	Latent heat of steam	587 cal g^{-1}

$$\text{Higher calorific Value} = \frac{(w + w_1)(t_2 - t_1)}{m} \text{ cal g}^{-1}$$

$$= \frac{(1500 + 470)(27.3 - 25.0)}{0.73}$$

$$= 6206.8 \text{ cal g}^{-1}$$

$$\frac{6206.8 \times 4.187}{10^{-3}} \text{ J kg}^{-1}$$

$$= \mathbf{25969.3 \times 10^3 \text{ J kg}^{-1}}$$

Net calorific value = HCV – heat released by the condensation of steam

$$= 25969.3 \times 10^3 - \frac{0.09 \times 25 \times 587 \times 4.187}{10^{-3}}$$

$$= \mathbf{25416.3 \times 10^{-3} \text{ J.kg}^{-1}}$$

5.6. PETROLEUM

In nature, there are two large reservoirs of organic compounds, namely petroleum and coal. Petroleum is the principal source of aliphatic organic compounds while coal is the main source of aromatic compounds. Thus coal, petroleum and natural gas are the major sources of energy. These non-renewable sources of energy (fossil fuels) are believed to have been formed millions of years ago by the slow decay of buried marine plants and animals deep down the earth surface by the action of high pressure and temperature. Petroleum, the only primary liquid fuel in nature, contains mainly a complex mixture of aliphatic hydrocarbons with small amount of other organic compounds containing nitrogen, oxygen and sulfur. The leading petroleum producing countries in the world are, Iran, Iraq, Kuwait and Saudi Arabia. America is the other major oil producing country in the world. India's petroleum demand is mainly met with imports since its production is meager.

In Biblical times, petroleum was used for coating bodies of boats, walls of temples, for embalming and other purposes. Just at the beginning of 20th century, kerosene was the chief product of petroleum which was used as fuel for illumination and heating. Half century ago, petrol (gasoline) was the gem of petroleum industry which was used as fuel for all automobiles and aero planes. Today, petroleum is considered to be a 'treasure' of industrial chemicals called petrochemicals.

The reserves of these fossil fuels have accumulated over millions of years and if we continue to consume them at the present rate, we will eventually exhaust the supply soon. According to estimation by National Academy of Sciences, Washington, the world supplies of natural gas, coal and petroleum will essentially be depleted by the first quarter of 21st century.

Petroleum refining (Fractionation)

Petroleum or crude oil is a dark colored, viscous oily liquid with an unpleasant smell. It is composed mainly of aliphatic hydrocarbons. It has to be processed before being used for diverse commercial purposes. It is done by a process called refining. The process of separation of crude oil into different useful fractions on the basis of their boiling points is called the petroleum refining or fractionation.

The refining is carried out by using an efficient bubble cap type fractionating column (Fig. 5.2)

The crude oil, freed from water and sulfur, is first heated in a preheater to 350 – 400°C whereby all volatile components are almost converted into vapors. The vapors are then fed into a specially designed large bubble cap fractionating column. The fractionating column permits separation of petroleum into a number of fractions having definite boiling point range.

The more volatile component condenses on the upper plates of the fractionating column while the less volatile fraction is collected on the lower plates. The various fractions condensed and collected at different parts of the column are continuously withdrawn and some of them are further fractionated and processed. The important fractions produced by refining of petroleum, their boiling point range and their applications are shown in Table 5.2.

5.7. CRACKING

Primary distillation of petroleum (crude oil) will never yield more than 20 percent of straight run gasoline. The demand for gasoline on the other hand, is increasing steeply. Accordingly, the straight run gasoline is found to be inadequate to meet the market demand. The present day gasoline demand is heavily supplemented by the '**Cracking**' process.

Cracking is defined as the process of breaking (degradation) of higher molecular weight hydrocarbons (high boiling) into lower molecular weight hydrocarbons (low boiling). Cracking in essence converts high boiling fractions for which there is little demand into high demand gasoline. Cracking process involves breaking (rupture) of carbon-carbon and carbon hydrogen bonds. It produces low boiling alkanes, alkenes and small amount of hydrogen.

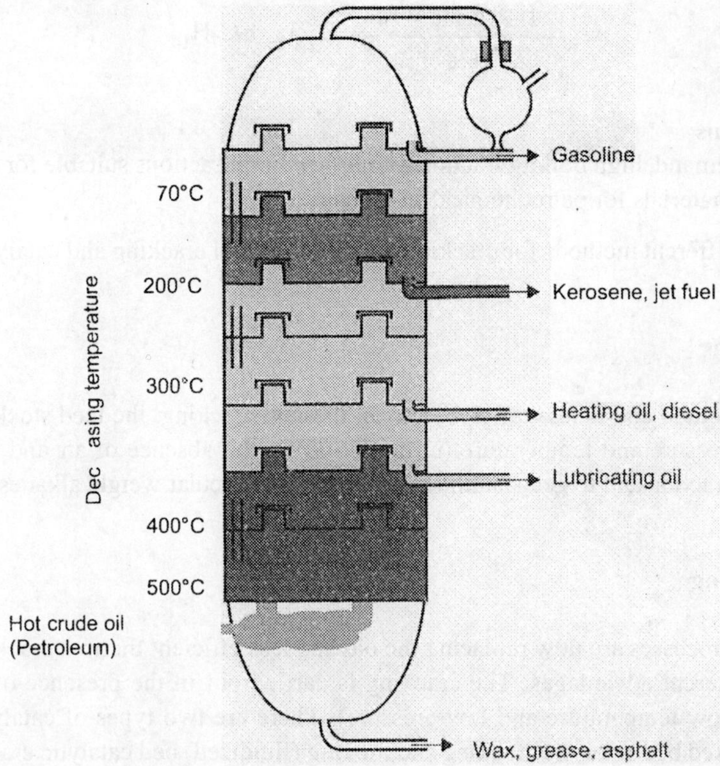

Fig. 5.2. Fractionation of petroleum (crude oil)

Table 5.2 Products of primary distillation of petroleum and their uses

Product fractions	*Boiling range* °C	*Composition of hydrocarbons*	*Uses*
Gas	Up to 20	$C_1 - C_4$	Domestic LPG fuel synthesis gas.
Petroleum Ether	20–70	$C_5 - C_6$	Fuel for petrol engines Aviation spirit
Kerosene	180–230	$C_{11} - C_{12}$	Domestic fuel for illumination and heating, jet fuel
Light gas oil	230–305	$C_{13} - C_{17}$	Fuel for diesel engines (Diesel) and furnaces
Heavy' gas oil	305–405	$C_{18} - C_{25}$	Fuel for generating (Lubricating oil) stations, lubricating oil, cracking stock.
Paraffin	Low melting	$C_{25} - C_{40}$	Waxes, candles, wax papers, matches
Solid residue (Asphalt)			Surfacing roads, Roofing, Petroleum coke.

$$C_{14}H_{30} \xrightarrow[\text{Absence of air}]{\text{Heat \& pressure}} C_7H_{16} + C_7H_{14}$$

Cracking process aims:

- To convert low demand, high boiling fraction into low boiling fractions suitable for automobiles.
- To produce raw materials for petrochemical industries.

There are two different methods for cracking, namely, thermal cracking and catalytic cracking.

(a) Thermal cracking

This is the earlier method and is less efficient one. In thermal cracking, the feed stock heavy oil is subjected to high pressure and temperature (up to 700°C) in the absence of air and catalyst. The cracked vapors are fractionated to get gasoline and other low molecular weight alkanes and alkenes.

(b) Catalytic cracking

Catalytic cracking processes are now replacing the old and less efficient thermal cracking methods because of their inherent advantages. The cracking is carried out in the presence of catalysts at milder conditions (low temperature and low pressure). There are two types of catalytic cracking methods, namely, fixed bed catalytic cracking and moving (fluidized) bed catalytic cracking. Of the two, the fluidized bed catalytic cracking method has some advantages over fixed bed cracking. Therefore fluidized bed catalytic cracking method is described here.

Fluidized (moving) bed catalytic cracking

In the fixed bed catalytic cracking process, the production has to be interrupted in order to regenerate the inactive catalyst. Interrupting of the production line being expensive, continuous processes have been developed in catalytic cracking where by regeneration of catalyst is carried out continuously without interfering with gasoline production. Fluidized bed catalytic cracking is one such method.

Principle

In fluidized bed catalytic cracking, the finely divided catalyst is kept agitated by gas streams (cracking fuel) so that it can be handled like a fluid system i.e., it can be pumped as a true liquid. There is a good contact between the catalyst and the reactant. Accordingly there is uninterrupted increased yield of products.

Optimum Conditions

Feed stocks	Gas oil, Heavy oil fractions
Catalyst used	$Al_2O_3 + SiO_2$ (Fluidized)
Temperature	550°C
Pressure	Little above the normal pressure
Production yield	10^6 gallons per day

A schematic diagram of fluidized bed catalytic cracking method is shown in Fig. 5.3

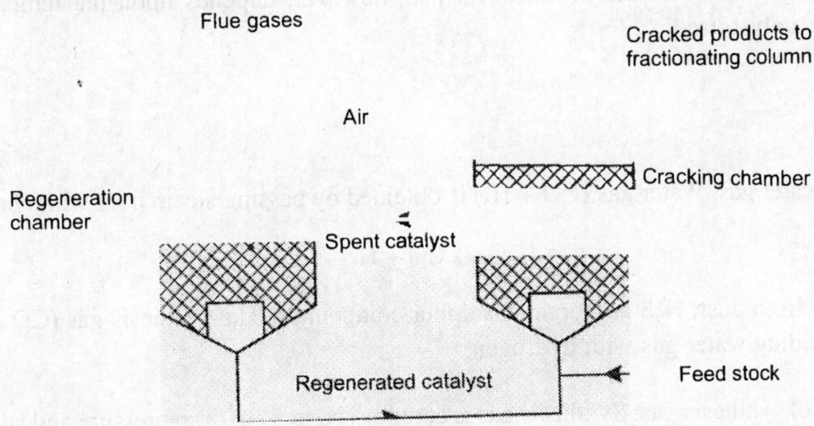

Fig. 5.3. Schematic diagram of fluidized bed catalytic cracking

Process

The finely divided catalyst bed is fluidized by the upward passage of feed stock vapors in a cracking chamber. Cracked vapors are withdrawn continuously from the top of the cracking chamber and directly fed into a fractionating column to separate into gases, gasoline, and uncracked oils. The uncracked oil may be cracked in a second stage cracking, there by increasing the overall yield of the cracked products.

Spent catalyst is drawn continuously from the bottom of the cracking chamber, transported in air stream to a regeneration chamber in which elemental carbon deposited on catalyst surface is burnt off. The regenerated catalyst is mixed with fresh feed stock and returned to the cracking chamber. The fluidized bed catalytic cracking, a prominent part of modern refinery, can be operated for periods measured in years before shut down is necessary.

5.8. SYNTHETIC PETROL

The principal elements present in coal are carbon and hydrogen. It is therefore not surprising that coal can be converted into liquid hydrocarbons, from which much needed petrol is obtained. This can be achieved either by the direct hydrogenation of coal or by the conversion of water gas obtained from the coal into petroleum. While Bergius process directly converts coal into petroleum, the Fischer-Tropsch process converts water gas into petroleum.

Fischer-Tropsch Process

Franz Fischer and Hans Tropsch showed that carbon monoxide and hydrogen in the presence of catalysts like Fe and Co at 180 - 250°C form a mixture of aliphatic hydrocarbons ranging from methane to wax. The nature of the products formed, however, depends upon the temperature, pressure and the catalyst used.

Process:

a) Production of water gas: Water gas ($CO + H_2$) is obtained by passing steam over white hot coal.

$$C + H_2O \rightarrow CO + H_2$$

Water gas is freed from dust, H_2S and organic sulphur compounds. The synthesis gas ($CO + 2H_2$) is obtained by blending water gas with hydrogen.

b) Hydrogenation of synthesis gas: Synthesis gas is compressed to 5 - 10 atm pressure and admitted into a catalytic reactor (Fig. 5.4). The catalyst, which is a mixture of cobalt (100 parts), thoria (5 parts) and magnesia (8 parts), is initially heated to 250°C. As a result of hydrogenation, saturated

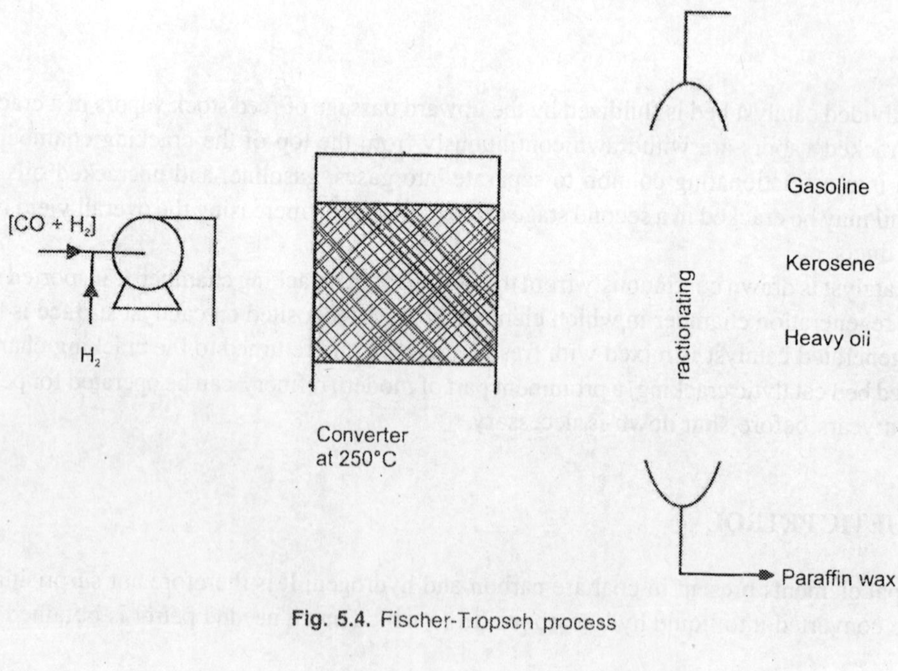

Fig. 5.4. Fischer-Tropsch process

and unsaturated hydrocarbons are produced. These products are produced through complex reactions and are exothermic.

$$nCO + 2nH_2 \qquad C_nH_{2n} + nH_2O$$

Unsaturated paraffins

$$nCO + (2n + 1)H_2 \rightarrow C_nH_{2n+2} + nH_2O$$

Saturated paraffins

The petroleum (mixture of hydrocarbons) so formed is next fractionated to yield petrol, diesel, kerosene, heavy oil and paraffin wax.

5.9. REFORMING OF PETROL

Gasoline used in automobiles mainly consists of a mixture of hydrocarbons – alkanes, cycloalkanes and aromatics. As a general rule, octane number of hydrocarbons increases as the hydrocarbons become more compact. It is found that branched chain hydrocarbons have higher octane numbers than the corresponding straight chain hydrocarbons. Aromatic hydrocarbons have higher octane numbers than non-aromatic hydrocarbons. Benzene, toluene, xylenes, for example, have octane numbers above 100.

Straight run gasoline obtained from primary distillation of crude oil contains mainly straight chain hydrocarbons. As such, it has a very low octane rating, not greater than 60. But internal combustion engines require gasoline of octane number above 90 in USA and 75–85 in India. It, therefore, follows that if the percentage of branched chain hydrocarbons and aromatic hydrocarbons can be increased, the octane rating will be improved. This improvement can be achieved by catalytic reforming process. The improved gasoline (increased in octane rating) is called the stabilized gasoline.

Catalytic-Reforming

Catalytic reforming is a process of upgrading gasoline (increasing its octane rating) in the presence of a catalyst. The increase in octane number of virgin gasoline occurs through structural modifications such as conversions of straight chain hydrocarbons into branched, cyclic and aromatics. All refineries have gasoline reforming units.

Reforming Process

Typical reaction conditions are:

Feed stocks	Virgin gasoline.
Catalyst	Platinum-supported on alumina silica base.
Temperature	470–525°C
Pressure	15–50 atmosphere.

The feed stocks are pretreated to remove sulfur and nitrogen content to acceptable levels to avoid platinum catalyst being poisoned. The vapors of feed stock are mixed with hydrogen and preheated to 500°C. The mixture is compressed and then fed into a series of three cylindrical reactors containing catalyst. The reformed products are fractionated to get stabilized gasoline.

Reforming reactions

The main reactions taking place during catalytic reforming process are:

- Isomerisation
- Dehydrogenation
- Cyclisation and dehydrogenation
- Hydrocracking

(i) Isomerisation: The straight-chain hydrocarbons are converted into branched chain hydrocarbons.

$$\underset{\text{n – heptane}}{CH_3-CH_2-CH_2-CH_2-CH_2-CH_2-CH_3} \rightarrow \underset{\text{2 – methyl hexane}}{CH_3-\underset{\underset{CH_3}{|}}{CH}-CH_2-CH_2-CH_2-CH_3}$$

(ii) Dehydrogenation: Cycloalkanes undergo dehydrogenation.

$$\underset{\text{Cyclohexane}}{C_6H_{12}} \rightarrow \underset{\text{Benzene}}{C_6H_6} + 3H_2$$

(iii) Cyclisation and dehydrogenation: Straight chain hydrocarbons undergo cyclisation followed by dehydrogenation to produce aromatic hydrocarbons.

$$\underset{\text{n—hexane}}{CH_3-CH_2-CH_2-CH_2-CH_2-CH_3} \rightarrow \underset{\text{cyclohexane}}{C_6H_{12}} + H_2$$

$$\underset{\text{Cyclohexane}}{C_6H_{12}} \rightarrow \underset{\text{Benzene}}{C_6H_6} + 3H_2$$

$$\underset{\text{n – heptane}}{CH_3-CH_2-CH_2-CH_2-CH_2-CH_2-CH_3} \rightarrow \underset{\text{Toluene}}{C_6H_5-CH_3} + 4H_2$$

(iv) Hydrocracking: n – Paraffins undergo hydrocracking (cracking in the presence of H_2 and catalyst) to produce light gases that are removed from gasoline to improve the octane number

$$\underset{\text{n – heptane}}{CH_3-(CH_2)_5-CH_3} + H_2 \rightarrow \underset{\text{propane}}{CH_3-CH_2-CH_3} + \underset{\text{n – butane}}{CH_3-CH_2-CH_2-CH_3}$$

5.10. KNOCKING

The efficiency of power production in spark ignited internal combustion (IC) engines is related to the compression ratio (CR). The CR is the ratio of the cylinder volume (V_1) at the bottom of its stroke (suction) to the cylinder volume (V_2) when the piston is at the top of its stroke (compression).

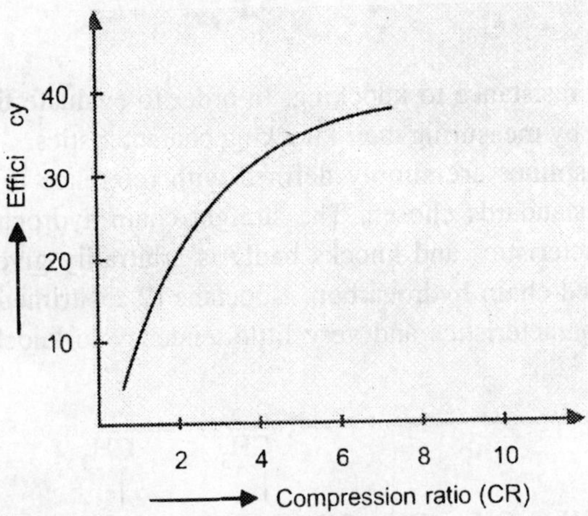

Fig. 5.5. Variation of IC engine efficiency with compression ratio

This ratio is always-greater than one since V_1 being greater than V_2. The efficiency of power output increases with the increase in CR. The variation of efficiency with CR is shown in Fig. 5.5.

The nature of he curve suggests that the efficiency of IC engine increases with compression ratio. Increased compression ratio however, brings benefits as well as problems.

In IC engines, the mixture of gasoline and air is drawn into the cylinder. The fuel-air mixture is compressed by the piston (compression stroke) and is ignited by an electric spark. As the flame front travels towards feed end of the combustion chamber, rapidly expanding combustion gases compress the remaining unburnt fuel ahead of flame front and raise its temperature. If the flame front travels rapidly at an optimum speed, the combustion of unburnt fuel takes place rapidly but smoothly. On the other hand, if the flame front travels too slowly, the entire last portion of the fuel mixture may get heated up beyond its ignition temperature and undergo instantaneous explosive combustion. This produces thermal shock waves which hit the cylinder walls and piston. This results in the emitting of characteristic rattling sound called **"knocking"** or **"pinking"**. The tendency of knocking increases with CR.

The following adverse effects are observed due to gasoline knock.

- It produces undesirable rattling noise.
- It increases the fuel consumption.
- It results in decreased power output.
- It causes mechanical damage due to overheating, to engine parts such as spark-plug, piston and engine walls.
- The driving becomes rather unpleasant.

Remedial Measures

The knocking menace in IC engines can be controlled and contained through following measures.

- By using high rating gasoline.
- By using critical compression ratio.
- By using anti-knocking agents.

Octane Number

The gasolines vary in their resistance to knocking. In order to evaluate their quality, an arbitrary octane scale was developed by measuring their knocking characteristics.

Octane numbers of gasolines are simply defined with reference to n-heptane (C_7H_{16}) and isooctane(C_8H_{18}), the two standards chosen. The straight chain hydrocarbon, n-heptane, which has poor combustion characteristics and knocks badly is arbitrarily given an octane number of 'zero' where as the branched chain hydrocarbon, isooctane (2,2,4-trimethyl pentane), which has an excellent combustion characteristics and very little tendency to knocking is given an octane number of 100.

$$CH_3-CH_2-CH_2-CH_2-CH_2-CH_2-CH_3; \quad \begin{array}{c} \quad CH_3 \qquad\quad CH_3 \\ \quad | \qquad\qquad\;\; | \\ CH_3-C-CH_2-CH-CH_3 \\ | \qquad\qquad\qquad \\ CH_3 \qquad\qquad\qquad \end{array}$$

n – heptane O.N.= 0.0 isooctane, O.N. = 100

The Octane number of a gasoline is "the percentage volume of iso-octane in a mixture of iso-octane and n-heptane blend, which has the same knocking characteristic as the gasoline sample under test." The different gasoline samples are rated by their octane numbers. Higher the octane number of a gasoline, least is its tendency for knocking, but higher is its combustion characteristics. 2,3-Dimethyl pentane has been assigned an octane number 75, because it has the knocking tendency as that of a mixture of n-heptane and iso-octane containing 75% by volume of iso-octane.

It has been found that knocking tendency is largely related to chemical structures of fuels. The decreasing tendency of fuels to knocking is as fallows.

Straight Chain > Branched Chain > Cycloalkanes > alkenes > Aromatics
Hydrocarbons alkanes

Antiknocking Agents

The use of high compression ratio in modern IC engines demands gasoline of high quality with least tendency of knocking. The octane rating of many fuels can be increased by the addition of certain organometallic compounds to gasoline (doping). The substances added to motor or aviation gasoline for controlling and containing the knocking in IC engines are called**'antiknocking'** agents.

The common commercial antiknocking agents used are:

(i) Tetraethyl lead (TEL) $Pb(C_2H_5)_4$
(ii) Tetramethyl lead (TME) $Pb(CH_3)_4$
(iii) Mixed methylethyl lead.

Among the antiknocks, TEL is perhaps the most widely used, since it is cheap and more effective in increasing the octane rating of fuels. The normal dosage of TEL is $4cm^3$ per gallon of gasoline. It is normally used along with ethylene dibromide or ethylene dichloride.

Table 5.3 Octane numbers of some hydrocarbons

Name	*Structure*	*Octane number*
n-Heptane	$CH_3 - (CH_2)_5 - CH_3$	0.00
3-Methyl hexane	$CH_3 - CH_2 - CH - CH_2 - CH_2 - CH_3$ | CH_3	52.00
3- Ethylpentane	$CH_3 - CH_2 - CH - CH_2 - CH_3$ $CH_2 - CH_3$	65.00
2,3- Dimethyl pentane	$CH_3 - CH - CH - CH_2 - CH_3$ | | CH_3 CH_3	75.00
2,2,4- Trimethyl pentane (iso-octane)	CH_3 $CH_3 - C - CH_2 - CH - CH_3$ | | CH_3 CH_3	100.00
Benzene		106.00
Toluene	$C_6H_5 - CH_3$	118.00

Octane no above 100 : 100 ml of iso-octane + x cm^3 TEL, O.N = 100 + x.

It is believed that during combustion of gasoline, TEL forms Pb and PbO. These species act as free radical chain inhibitors and thus curtail the propagation of explosive chain reaction and thereby minimizing knocking. However, if TEL is used, the species Pb and PbO may get deposited on engine parts and cause mechanical damage. In order to minimize the damage to engine parts, TEL is always used along with ethylene dibromide or chloride. The function of these halogen compounds is to convert the less volatile Pb and PbO into more volatile $PbBr_2$ or $PbCl_2$ which escape into air along with exhaust gases.

Unleaded Petrol

In order to enhance the octane rating of gasoline in IC engines, gasoline is normally, mixed with 0.1 percent TEL with gasoline. This increases octane rating by 10–15 points. Unfortunately mixing of TEL with gasoline brought advantages together with disadvantages. In order to overcome the harmful effects of TEL, it is slowly being phased out from gasoline. The gasoline (petrol) free from lead is called **'unleaded petrol'** Of late, it has become mandatory to use unleaded petrol for automobiles.

The compelling reasons for using unleaded petrol are explained as under:

- Lead is toxic. The recognition of health hazard associated with its release into atmosphere from automobile exhaust lead to mandatory phasing out of TEL in gasoline.
- Further factor for reducing TEL is the requirement of vehicles to be fitted with catalytic converters to control pollutants. Only unleaded petrol can be used for vehicles fitted with catalytic converters because lead inactivates the catalysts used in converters.

Unleaded petrol rating can be enhanced by the addition of certain compounds in place of TEL. The octane rating enhancers such as methyl-t-butyl ether (MTBE), Ethyl-t-butyl ether (ETBE), methanol, ethanol which have octane rating above 100, are added to unleaded petrol in place of TEL.

5.11. DIESEL KNOCKING

The diesel engine is a compression ignition engine which uses long straight-chain hydrocarbon fuels having boiling range 180–360°C.

In the case of diesel engine, air is passed into the cylinder and compressed to about 500psi (upstroke). As a result of compression, the temperature of air rises to about 500°C. The diesel oil is injected towards the end of compression stroke in the form of fine droplets into the compressed air in the cylinder heated to 500°C by compression. The injected oil droplets absorb the heat from the heated air, get vaporized, attain self ignition temperature and burn spontaneously during downward stroke. However, the combustion of fuel in diesel engines is not instantaneous but there exists an interval of time between fuel injection and start of ignition. This interval of time is called **'ignition delay''** or **'ignition lag'**. This is an important parameter of diesel fuels and diesel knock. If the fuel has a short ignition delay, the fuel injected into the burning mixture continues to burn at the rate at which it is injected. On the other hand, if the fuel has a long ignition delay; then the fuel accumulation occurs in the engine even before ignition. When ignited, an explosive combustion occurs with sudden increase in pressure. This is called **'diesel knock.'** Longer the ignition delay, larger is the diesel knock.

The advantages of using diesel fuels are:

- The diesel fuel is cheaper than petrol.
- The fuel consumption per unit power generation is less.
- The thermal efficiency is high.
- The exhaust contains fewer pollutants.

The diesel knock can be minimized by using fuels having following characteristics.

- The fuels should have straight chain structure.
- The fuels must have a short ignition delay.
- The fuel ignition temperature should be less than that of compressed air.
- The fuel used must have high cetane rating
- The cetane rating of diesel fuel can be raised by additives such as ethyl nitrate ($CH_3 - CH_2 - O - NO_2$), amyl nitrate ($CH_3 - CH_2 - CH_2 - CH_2 - CH_2 - O - NO_2$), etc.

Cetane Number

The cetane number is a measure of the ease with which the given diesel fuel will undergo compression ignition. The cetane rating for diesel fuels is equivalent to octane rating for gasoline.

Two hydrocarbons namely α-methyl naphthalene (aromatic) and n-cetane (aliphatic straight chain) are specified as standards. Since n-cetane has low ignition lag, its cetane number is arbitrarily fixed as 100, while α-methyl naphthalene has long ignition lag and its cetane number is fixed as zero.

$C_{10}H_7-CH_3$	$CH_3-(-CH_2-)_{14}-CH_3$
α-Methylnaphthalene	n – Cetane
Cetane number 0	Cetane number 100.

The cetane number of a diesel fuel is the percentage of n-cetane in a mixture of n-cetane and α-methyl naphthalene giving the same performance as that of diesel fuel under test.

Thus the structural requirements for a diesel fuel are precisely those which have to be avoided in gasoline.

- The straight chain alkanes like n-cetane, which ignite readily are good diesel fuels.
- Aromatics like α – methyl naphthalene which has long ignition delay is poor diesel fuels.
- Cetane numbers in practice range from about 25–50, increasing with increasing engine speed.
- Certain additives can be used to raise the cetane ratings. Alkyl nitrates such as ethyl nitrate $(CH_3-CH_2-O-NO_2)$ and amyl nitrate $(CH_3-CH_2-CH_2-CH_2-CH_2-O-NO_2)$ are the common additives used to raise diesel rating.

5.12. CATALYTIC CONVERTER

The catalytic converter was one of the greatest emission control inventions in the history of monitoring carbon and greenhouse gas emissions. A catalytic converter is a device used to reduce the toxicity of emissions from an internal combustion engine. Most modern cars are equipped with three-way catalytic converters. "Three-way" refers to the three regulated emissions it helps to reduce carbon monoxide, unburnt hydrocarbons and nitrogen oxide molecules. A three-way catalytic converter has three simultaneous tasks:

1. Reduction of nitrogen oxides to nitrogen and oxygen:

$$2NO_x \rightarrow xO_2 + N_2$$

2. Oxidation of carbon monoxide to carbon dioxide:

$$2CO + O_2 \rightarrow 2CO_2$$

3. Oxidation of unburnt hydrocarbon (HC) to carbon dioxide and water:

$$C_xH_{2x+2} + [(3x+1)/2]O_2 \rightarrow xCO_2 + (x+1)H_2O$$

The catajytic converter consists of a core, or substrate, which is often a ceramic honeycomb, but stainless steel foil honeycombs are also used. The honey-comb surface increases the amount of surface area available to support the catalyst, and therefore is often called a "catalyst support". The core is added with a mixture of silica and alumina, which form a rough, irregular surface, with a far greater surface area than the flat core surfaces do. The catalyst is added to the mixture of silica and alumina suspension before being applied to the core.

In the catalytic converter, there are two different types of catalyst at work, a reduction catalyst and an oxidation catalyst. The reduction catalyst is the first stage of the catalytic converter and uses platinum and rhodium to help reduce the nitrogen oxide emissions. When such molecules come in contact with the catalyst, the catalyst converts them into O_2 and N_2. The oxidation catalyst is the second stage of the catalytic converter. It reduces the unburned hydrocarbons and carbon monoxide by burning (oxidizing) them over a platinum and palladium catalyst.

5.13. POWER ALCOHOL

When ethyl alcohol is used as an additive to motor fuels to act as a fuel for internal combustion engines, it is called power alcohol. Blends containing up to 25% of alcohol with petrol are used.

Manufacture of Power Alcohol

The important raw materials for the manufacture of power alcohol or ethyl alcohol are saccharine materials (such as molasses, sugar beets, sugar cane, etc), starchy materials (such as starch, potatoes, cereal grains, etc), cellulose materials (such as sulphite liquor from paper mills) and hydrocarbon gases.

Ethyl alcohol from molasses

Molasses is a dark colored viscous liquid left after the crystallization of cane sugar from cane juice. It contains 50–55% total sugar, of which 35–40% is cane sugar or sucrose. Molasses is converted into ethyl alcohol by means of yeast, which contains enzymes invertase and zymase responsible for fermentation.

$$\underset{}{C_{12}H_{22}O_{11}} \xrightarrow{\text{Invertase}} \underset{\text{Glucose}}{C_6H_{12}O_6} + \underset{\text{Fructose}}{C_6H_{12}O_6}$$

$$\underset{\text{Glucose, Fructose}}{C_6H_{12}O_6} \xrightarrow{\text{Zymase}} 2C_2H_5OH + 2CO_2$$

Before mixing with yeast, the molasses is diluted with water to bring down the concentration of sugar to 10–12%. pH is kept between 4 and 5 to favor the function of enzymes, by using sulfuric acid. After fermentation for 48–60 hours, it is distilled to get ethyl alcohol which is mixed with water. Repeated distillation and condenstion can raise the alcohol content to 97.6%.

Absolute alcohol can be obtained from the above by azeotropic process in which benzene is used to form ternary mixture with water and alcohol. The ternary, constant boiling mixture boils at

65°C and pure dehydrated alcohol, which is absolute alcohol, boils at 78.5°C. So the vapors of ternary mixture go out at 65°C taking away all the water and leaving behind absolute alcohol.

Power alcohol as a fuel

Power alcohol is used as a fuel by blending with petrol in internal combustion engines as a motor spirit. Blends containing up to 25% of alcohol with petrol are used. Industrial alcohol containing 95% alcohol and 5% of water can also be blended with petrol but by using some blending agent such as benzene, ether, etc. Without the use of blending agent, industrial alcohol does not mix with petrol.

Advantages of alcohol-blended petrol

- Alcohol has an octane number of 90; while petrol has octane number of 60–70. Addition of power alcohol to petrol increases the octane number. Hence alcohol blended petrol possesses better anti-knock properties.
- Because of the higher octane number, alcohol-blended petrol can be used in engines with higher compression ratio. This compensates for the lower heating value of alcohol in the blend.
- There are no starting difficulties with alcohol-petrol blend.
- Lubrication in case of alcohol-petrol blend and pure petrol is the same.
- Air required for complete combustion is less.

isadvantages of alcohol-blended petrol

- Alcohol lowers the calorific value of petrol.
- Because of its considerable surface tension. it is difficult to atomize alcohol, particularly at low temperature. Hence, specific arrangements for starting alcohol-petrol-blend in carburettor are to be provided.
- Alcohol is easily oxidized to acids. Hence alcohol may cause corrosion.
- Air entering the cylinder needs to be regulated by modifying air inlet as less air is required for combustion.
- Alcohol absorbs moisture and as a result separation of alcohol and petrol layers takes place especially at low temperature. To avoid this, blending agents such as benzene or tol iene are used.

5.14. BIODIESEL

Biodiesel is an alternative fuel which can be made using a simple chemical process from waste vegetable oil. Biodiesel refers to a vegetable oil- or animal fat-based diesel fuel consisting of long-chain alkyl (methyl, propyl or ethyl) esters. Biodiesel is typically made by chemically reacting vegetable oils or animal fat with an alcohol. Biodiesel can be used alone, or blended with petrodiesel. It is a green fuel, does not contribute to the carbon dioxide (CO_2) burden and produces drastically reduced engine emissions. It is non-toxic and biodegradable.

Biodiesel is commonly produced by the trans-esterification of the vegetable oil or animal fat feedstock. This process involves combining any natural oil (vegetable or animal) with virtually any alcohol, and a catalyst. The most commonly used alcohol is methanol to produce methyl esters (commonly referred to as Fatty Acid Methyl Ester—FAME) as it is the cheapest alcohol available. Ethanol can be used to produce an ethyl ester (commonly referred to as Fatty Acid Ethyl Ester—FAEE) biodiesel.

$$\begin{array}{l} CH_2COOR \\ | \\ CHCOOR \\ | \\ CH_2COOR \end{array} + 3CH_3OH \overset{\text{catalyst}}{\Leftrightarrow} 3CH_3COOR_x + \begin{array}{l} CH_2OH \\ | \\ CHOH \\ | \\ CH_2OH \end{array}$$

Vegetable oil   *Biodiesel*   *Glycerin*

Biodiesel has combustion properties very similar to those of petroleum diesel, and can replace it in most current uses.

A variety of oils can be used to produce biodiesel. These include:

- Virgin oil feedstock: rapeseed and soybean oils are most commonly used, others can be jatropha and other crops such as mustard, flax, sunflower, palm oil, coconut, etc.
- Waste vegetable oil (WVO).
- Animal fats including tallow, lard, yellow grease, chicken fat, and the by-products of the production of fatty acids from fish oil.
- Algae, which can be grown using waste materials such as sewage and without displacing land currently used for food production.

Advantages

- Renewable: Biodiesel is derived from vegetable oil which is essentially grown—a sustainable resource that will not run out.
- Carbon Neutral—Biodiesel use does not lead to any overall change in the amount of CO_2 in the atmosphere. The vegetables from which the oil has been extracted remove CO_2 from the atmosphere to grow. When biodiesel is burned the CO_2 is released back into atmosphere.
- Less noxious, non-toxic—Biodiesel lacks the unpleasant odour of petroleum diesel. Users can expect a near 100% reduction in sulphur dioxide (SO_2), 40-60% reduction in soot and particulates, 10-50% reduction in CO, and a reduction in all poly aromatic hydrocarbons.
- Unlike petroleum diesel, it is biodegradable.
- Simple to make, and can be produced from waste vegetable oil.
- Classed as non-hazardous because it is non-toxic and has a high flash-point. Safer to store and transport.

- Burns more efficiently than petroleum diesel.
- Substantially higher lubricity means it can reduce engine wear and hence prolong engine life

.1.. BIOGAS

Biogas typically refers to a mixture of gases produced by the anaerobic digestion or fermentation of organic matter. The organic matter can be manure, sewage sludge, municipal solid waste, biodegradable waste or any other biodegradable feedstock such as plant materials and crops. Biogas is mainly methane and carbon dioxide. It is also referred to as biofuel, sewerage gas, Klar gas, sludge gas, will-o-the wisp of marsh lands, fool's fire, *gobar* gas (cow dung gas).

A **biogas plant** is the anaerobic digester that treats biodegradable wastes to produce biogas. These plants can be fed with energy crops such as maize silage or biodegradable wastes including sewage sludge and food waste. During the process, an air-tight tank transforms biomass waste into methane, producing renewable energy that can be used for heating, electricity, and many other operations that use an internal combustion engine. A simple schematic representation of a biogas plant is shown in Fig. 5.6.

Anaerobic digestion is carried out in an air tight cylindrical tank which is known as digester. A digester is made up of concrete bricks and cement or steel. It has a side opening (charge pit) into which organic materials for digestion are incorporated. The cylindrical container above the digester collects the gas. Digested sludge is removed from the basin and is used as fertilizer. Usually digesters are buried in soil in order to benefit from insulation provided by soil. In cold climate, digester can be heated by the installations provided from composting for the agricultural wastes.

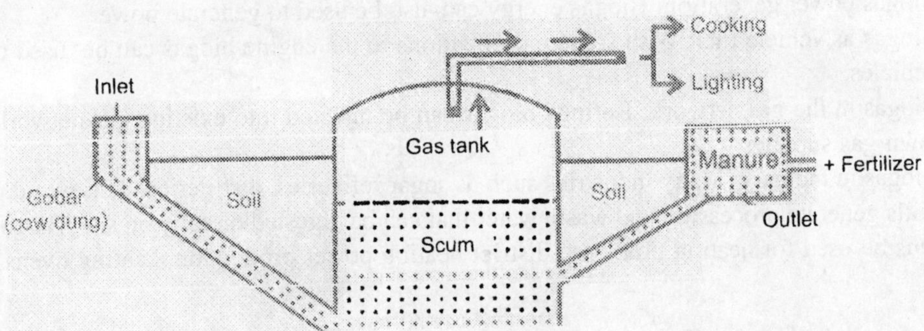

Fig. 5.6: Biogas plant

Typical composition of biogas

Compound	CH_4	CO_2	N_2	H_2	H_2S
(%)	50–75	25–50	0–10	0–1	0–3

Biogas upgrading

Raw biogas produced from digestion is not of high quality to be used as fuel gas for machinery. The corrosive nature of H_2S alone is enough to destroy the internals of a plant. Methane in biogas can be concentrated via a biogas upgrader to the same standards as fossil natural gas, by removing carbon dioxide, water, hydrogen sulfide, and particulates, if present.The most commonly used method is water washing where high pressure gas flows into a column where the carbon dioxide and other trace constituents are scrubbed by cascading water running counter-flow to the gas.

Uses

It may also be used in fueling internal combustion engines and fuel cells for production of mechanical work and/or electricity. If cleaned up to adequate standards it may be injected into gas pipelines and provide illumination and steam production. Finally, through a catalytic chemical oxidation methane can be used in the production of methanol production.

- Biogas for heating: Biogas can be used readily for heating, cooking, space and water heating, drying.
- Biogas power generation: Biogas energy can also be used to generate power.
- Biogas as vehicle fuel: With some modifications to the engine biogas can be used to run vehicles.
- Biogas in the gas network: Refined biogas can be injected into existing gas networks for town gas supplies.
- Biogas in industry: Many industries such as sugar refineries, distilleries, dairies and paper mills generate processing and waste water that can be digested directly on site. Biogas can thus be used for heating premises, district heating power production, heating ovens, etc.

Advantages

- Renewable and increased energy security. Biogas offsets non-renewable resources, such as coal, oil, and fossil fuel-derived natural gas.
- High nutrient fertilizer produced in excess. Produces enriched organic manure, which can supplement or even replace chemical fertilizers.
- Environment friendly. Leads to improvement in the environment, and sanitation and hygiene. Reduces methane and carbon dioxide release into the air.

- Household wastes and bio-wastes can be disposed of usefully and in a healthy manner.
- Anaerobic digestion inactivates pathogens and parasites, and is quite effective in reducing the incidence of water-borne diseases.
- Cheap to produce. Any biodegradable matter can be used as substrate.
- Reduces landfill sites, sewage drainage, and farm manure.
- Clean/ quiet fuel for cars and trucks.

Disadvantages

- Reduced power from vehicles.
- Extra time required to produce in comparison to other energy resources.
- The process of using biogas on a large scale is not economically viable and it is very difficult to enhance the efficiency of biogas systems.
- It is also somewhat unstable, making it prone to explosions if the methane comes in contact with oxygen and becomes flammable in nature.

REVIEW QUESTIONS

I. Justify the following statements:

1. Gross calorific value of a fuel is always higher than its net calorific value.
2. Combustion of chemicals releases heat energy.
3. TEL is always used along with ethylene dibromide, as an antiknocking agent.

II. Answer the following questions:

1. What are chemical fuels? How are they classified? Give examples for each class.
2. Define: (a) Calorific value (b) Gross calorific value (c) Net calorific value.
3. What is petroleum? Give an account of petroleum refining.
4. Name the important fractions obtained in petroleum refining. Mention their applications.
5. What is meant by cracking? Describe with a neat diagram, fluidized bed cracking method.
6. What is meant by reforming? Explain the principal reactions occurring during reforming process.
7. Explain the phenomenon of knocking in gasoline engines. How *is* it prevented?
8. Define: (a) Octane number, (b) Cetane number.
9. Explain the phenomenon of knocking in diesel engines.
10. How calorific value of a fuel is determined using bomb calorimeter?
11. What is power alcohol? What are the advantages and disadvantages of blending alcohol with petrol?
12. What is biodiesel? How is it prepared? What are the advantages?
13. What are the functions of a catalytic converter?
14. Explain Fischer-Tropsch method of synthesizing petrol.
15. Give an account of biogas as a fuel.

CHAPTER 6

Solar Energy

6.1. INTRODUCTION

Survival has always been the main preoccupation of mankind. For many thousands of years, food, shelter and protection against harsh weather and wild animals were the primary requirements of early man. His needs were a few and the means of fulfilling the needs were also simple. It is estimated that early man had a daily energy consumption rate of around 2500 kilocalories. Over the years, early man incorporated new options to fulfill the basic requirements of survival, and at the same time altered the energy cycles of primitive society. A time came when technology was developed to simplify everyday productive activities. In the quest of more comfort in life, man went behind the facilities developed and adopted a more complex life styles. Along with the dawn of progress, came larger and larger requirements for energy. Modern life became an endless chain of activities and events, demanding more and more energy. The per capita consumption of energy rose to over 2,30,000 kilocalories per day, two orders of magnitude larger than that of primitive man.

Now, the energy has joined the rank of food, shelter and clothing – three essential commodities of mankind. Till the middle of 19th century, wood and coal constituted the main sources of energy with smaller contributions from petroleum. In the last quarter of the 20th century, the dominant energy sources switched to petroleum, natural gas, hydro power and nuclear energy. As the reserves of fossil fuel are very limited, and are being depleted very fast, search for alternate sources energy has gained a lot of importance. The world is looking towards natural resources, such as solar energy, wind energy, bioenergy, ocean energy, geothermal energy, etc. Out of these, solar energy has the greatest potential. In the last few decades, lot of effort has gone in converting and utilizing solar energy in different forms.

6.2. SOLAR ENERGY UTILIZATION AND CONVERSION

A wide range of power technologies exist which can make use of the solar energy reaching earth and converting them into different useful forms of energy. Solar energy utilization can be of two types – **direct solar power** and **indirect solar power.**

Direct solar power

Direct solar power involves only one step transformation into a usable form.

Examples:

- Sunlight hits a photovoltaic cell (also called a photoelectric cell) generating electricity.
- Sunlight hits the dark absorber surface of a solar thermal collector and the surface warms. The heat energy may be carried away by a fluid circuit.
- Sunlight strikes a solar sail on a space craft and is converted directly into a force on the sail which causes motion of the craft.

Indirect solar power

Indirect solar power involves more than one transformation to reach a usable form.

Examples:

- Vegetation uses photosynthesis to convert solar energy to chemical energy, which can later be burnt as fuel to generate electricity (biofuel). Methane (natural gas) may be derived from the biofuel.
- Hydroelectric dams and wind turbines are powered by solar energy through its interaction with the Earth's atmosphere and the resulting weather phenomena.
- Ocean thermal energy production uses the thermal gradients that are present across ocean depths to generate power. These temperature differences are ultimately due to the energy of the sun.
- Energy obtained from oil, coal, and peat originated as solar energy captured by vegetation in the remote geological past and fossilized. Hence the term fossil fuel. The great time delay between the input of the solar energy and its recovery means these are not practically renewable and therefore not normally classified as solar power.

Advantages

- Solar power is relatively pollution free.
- Facilities can operate with little maintenance or intervention after initial setup.
- Solar power is becoming more and more economical as costs associated with production decreases, and the technology becomes more effective in energy conversion.
- Solar power can be viewed as a local resource because of regional climatic variances.
- Some countries, regions, etc (such as island communities, desolate regions and ocean-going vessels) are harvesting solar power as a viable energy resource than purchasing energy from other costly sources.

Disadvantages

Solar power at the Earth's surface has a number of disadvantages which must be addressed as engineering problems before it can become an effective source of energy supply:

- It is only practical in certain areas with a favorable climate and latitude. That is, areas near the tropics and which are relatively cloud free.
- It is not available at night and is reduced when there is cloud cover, decreasing the reliability of peak output performance.
- It must be converted into some other form of energy to be stored for times when conditions are prohibitive or to drive transport.
- Solar cell technologies produce DC power which must be converted to the AC power when used in distribution grids.

6.3. PHOTOVOLTAIC CELLS

Introduction

Photovoltaic cells orsolar cells as they are often referred to are semiconductor devices that convert sunlight into direct current electricity. As long as light is shining on the solar cell, it generates electrical power. When the light stops, electricity stops. Solar cells never need recharging like a battery. Some have been in continuous outdoor operation on earth or space for over 30 years.

Groups of photovoltaic cells are electrically configured into modules and arrays, which can be used to charge batteries, operate motors, and to power any number of electrical loads. With the appropriate power conversion equipment, photovoltaic systems can produce alternating current (AC) compatible with any conventional appliances.

Photovoltaic cell is based on photoelectric effect. The photoelectric effect was first noted by a French physicist, Edmund Bequerel, in 1839. He found that certain materials would produce small amount of current when exposed to light. In 1883, Fritts (US) developed the first large area solar cell using Se film. In 1905, Albert Einstein described the nature of light and the photoelectric effect on which photoelectric technology is based, for which he later won a Noble prize in physics. In 1954, the first 6% efficient solar cells were developed by Bell Laboratories. In the 1960s, the space industry began to make the first serious use of the technology to provide power aboard spacecraft. Through the space programs, the technology advanced, its reliability was established, and the cost began to decline. During the energy crisis in the 1970s, photovoltaic technology gained recognition as a source of power for non space applications.

Photovoltaics is the technological symbol for a future sustainable energy supply system. A considerable amount of money is invested in research, development and demonstration.

Working of a photovoltaic cell

Semiconductors have the capacity to absorb light and to deliver a portion of the energy of the absorbed photons to carriers of electrical current – electrons and holes. A semiconductor diode separates and collects the carriers and conducts the generated electrical current preferentially in a specific direction. Thus solar cell is a semiconductor diode that has been carefully designed and constructed to efficiently absorb and convert light energy from the sun into electrical energy. A conventional solar cell structure is shown in Fig 6.1.

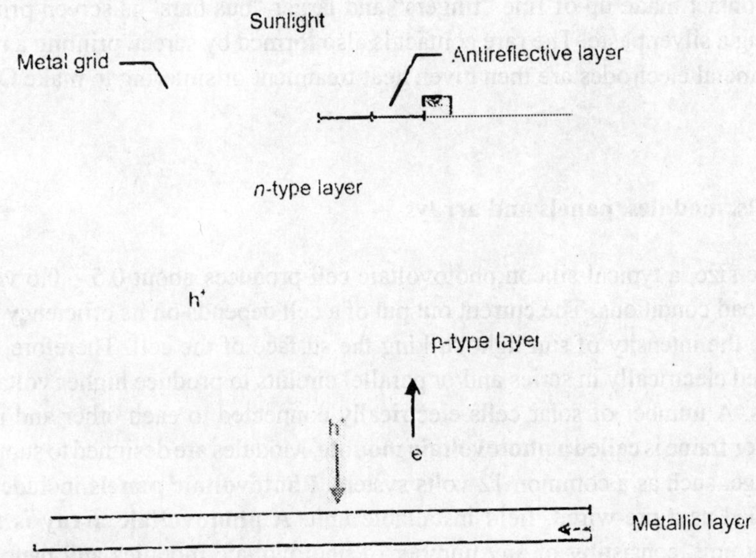

Fig. 6.1. Solar cell structure

A typical silicon photovoltaic cell is composed of a thin wafer consisting of an ultra thin layer of phosphorous doped (n-type) silicon on top of boron doped (p-type) silicon. Hence a p-n junction is formed between the two. A metallic grid forms one of the electrical contacts of the diode and allows light to fall on the semiconductor between the grid lines. An anti reflective layer between the grid lines increases the amount of light transmitted to the semiconductor. The cell's other electrical contact is formed by a metallic layer on the back of the solar cell.

When light radiation falls on the p-n junction diode, electron – hole pairs are generated by the absorption of the radiation. The electrons are drifted to and collected at the n- type end and the holes are drifted to and collected at the p- type end. When these two ends are electrically connected through a conductor, there is a flow of current between the two ends through the external circuit. Thus photoelectric current is produced and available for use.

Construction of a solar cell

Poly crystalline silicon wafers are made by wire-sawing block-cast silicon ingots into very thin slices or wafers (250–350 μm). The wafer is etched slightly to remove the saw damage and to condition the surface for better light absorption. The wafers are usually lightly p-doped. To make a solar cell from the wafer, a surface diffusion of n-type dopants is performed on the front side of the wafer. This forms a p-n junction of a few hundred nanometers below the surface.

Antireflection coating of silicon nitride or titanium dioxide is then applied in a layer, several hundred nanometers thick, using plasma enhanced chemical vapor deposition (PEVCD) technique. The wafer is then metallized, whereby a full area metal contact is made on the back surface, and a grid like metal contact made up of fine "fingers" and larger "bus bars" is screen printed onto the front surface using a silver paste. The rare contact is also formed by screen printing a metal paste of aluminium. The metal electrodes are then given heat treatment or sintering to make Ohmic contact with the silicon.

Photovoltaic cells, modules, panels and arrays

Regardless of the size, a typical silicon photovoltaic cell produces about 0.5 – 0.6 volt DC under open circuit, no load conditions. The current out put of a cell depends on its efficiency and size, and is proportional to the intensity of sun light striking the surface of the cell. Therefore, photovoltaic cells are connected electrically in series and/or parallel circuits to produce higher voltages, currents and power levels. A number of solar cells electrically connected to each other and mounted in a support structure or frame is called **a photovoltaic module**. Modules are designed to supply electricity at a certain voltage, such as a common 12 volts system. **Photovoltaic panels** include one or more modules assembled as a pre-wired, field installable unit. A **photovoltaic array** is the complete power generating unit, consisting of any number of photovoltaic modules and panels. These are depicted in Fig. 6.2.

The performance of photovoltaic modules and arrays are rated according to their maximum DC power output under **Standard Test Conditions (STC).** Standard operating conditions are the operating temperature of 25°C and incident solar irradiance level of 1000 W / m^2 and under Air Mass 1.5 spectral distribution. Since these conditions are not always typical in the field, actual performance is usually 85 to 90 percent of the STC rating.

Importance of photovoltaic cells

- With the conventional energy reserves are fast depleting, photovoltaics is considered to be the future sustainable energy system. Solar energy, being unlimited, inexhaustible and renewable, photovoltaic cells are the means to harness this continuous energy source.
- Photovoltaic cells can serve for both off-grid and on-grid applications. It can be used for off grid professional devices and supply systems such as telecommunication equipment, solar

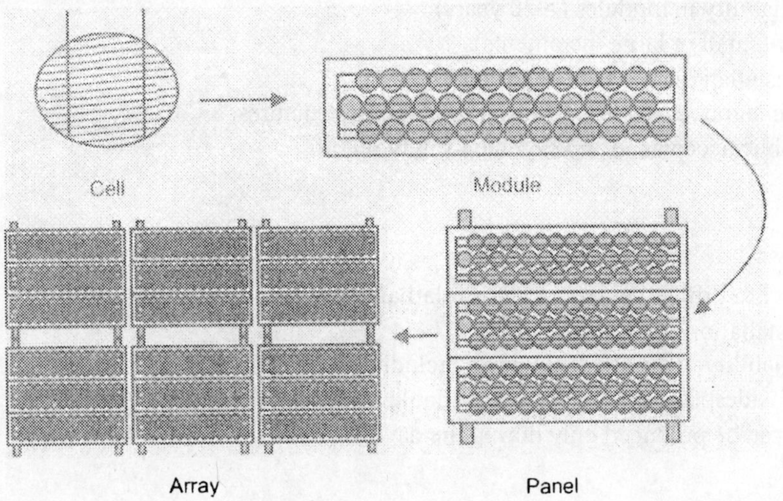

Fig. 6.2. Photovoltaic cells, modules, panels and arrays

home systems, etc. It can also be for large scale electricity generation as a substitute for and a complement to today's non sustainable energy processes.

- Photovoltaic energy conversion complies with the requirements of environmentally benign production schemes. During the operation, there is no harmful emission or transformation of matter (generation of pollutants), nor any production of noise or other by-products.
- Photovoltaic energy conversion is highly modular. This is important with respect to the development of electricity supply systems in many rural and remote areas, where grid extension is economically not feasible.
- The photovoltaic electricity can contribute considerably to the abatement of the man made global warming due to carbon dioxide.
- Photovoltaics can be used as roof integrated systems, providing power and also serving as optical shading elements for the space below and preventing overheating in the summer.
- Photovoltaic cells provide power for spacecraft and satellites, an extraterrestrial dimension of photovoltaics.
- Developments in the field of photovoltaic cells will boost the semiconductor industry and storage battery industries.

Advantages

- Energy source is vast and essentially infinite
- No emissions, no combustion or radioactive residues for disposal. Does not contribute to global climate change or pollution.
- Low operating costs (no fuel).
- No moving parts and no wear and tear.
- Ambient temperature operation (no high temperature corrosion or safety issues).

- High reliability in modules (> 20 years).
- Modular (small or large increments).
- Quick installation.
- Can be integrated into new or existing building structures.
- High public acceptance and excellent safety record.

Disadvantages

- Sunlight is a diffuse source, i.e., it is relatively low density energy.
- High installation costs.
- Poor reliability of auxiliary elements including storage.
- Lack of widespread commercially available system integration and installation.
- Energy can be produced only during the day time.

6.4. SILICON

By far the most common material used for solar cells is crystalline silicon, with multicrystalline silicon taking the lion's share. Crystalline silicon solar cells come in three primary categories.

(i) Single crystalline or monocrystalline wafers
(ii) Poly or multi crystalline wafers
(iii) Ribbon silicon – drawn from molten silicon, having a multicrystalline structure.

Silicon is the second member in the group IVA in the periodic table. It never occurs free in the nature, but in combination with oxygen forming oxides and silicates. Most of the earth's crust is made of silica and silicates. Silicon constitutes 26% of the earth's crust and is the second most abundant element in weight, next to oxygen.

Physical properties of silicon relevant to photovoltaics

Silicon is a semiconductor with a band gap of 1.12 eV at 25°C. At atmospheric pressure, silicon crystallizes into a diamond cubic structure. Vapour deposition below 500°C results in amorphous silicon. If reheated above this temperature, crystallization occurs. Unlike most of the compounds and elements, silicon contracts when melting and expands when solidifying.

Impurities incorporated in the silicon lattice during the crystal growth or during the post treatment ionize at low temperature, thus providing either free electrons or holes. Impurities from VA group are n-dopants and impurities from IIIA group are p-dopants. Phosphorous and boron represent these groups and are used in photovoltaic processing to control the semiconductor properties of silicon. Presence of transitional metals, Fe, Cr, Ni, degrade the minority carrier life time and the solar cell performance.

The relatively high refractive index limits the optical applications of silicon. The absorption and transmission properties in the 0.4–0.5 μm wavelength apectra are important in the performance of photovoltaic cells. In photovoltaic applications, silicon applied with antireflective layers is generally used. Silicon is brittle even when alloyed with small quantities of impurities. Shaping for photovoltaic applications require sawing and grinding. These mechanical operations are similar to those applied to glasses.

Chemical properties relevant to photovoltaics

Silicon is stable in the tetravalent state and has a strong affinity for oxygen, forming stable oxides and silicates. Elemental silicon readily oxidizes, forming a thin protective film of silica (SiO_2).

Silicon and carbon form a strong Si – C bond and stable products. Silicon carbide also finds various applications in photovoltaics and electronics. Primary uses exploit the abrasive properties of SiC for wafering silicon crystals. Silicon forms hydrides, and monosilane (SiH_4) is a key chemical compound for the production of amorphous silicon and the purification of silicon to semiconductor grade.

The chemical reactivity of silicon with chlorine is also very important. Trichlorosilane and tetrachlorosilane are both the intermediates and the by-products of the purification processes in upgrading metallurgical grade silicon to semiconductor grade, as these compounds are volatile at low temperature and can be decomposed to elemental silicon at higher temperature. Other chlorosilanes or halogenosilanes are also used in chemical vapour deposition applications. Silicon and germanium are isomorphous and mutually soluble in all proportions.

Solar grade silicon

Elemental silicon is used in photovoltaic cells as the main semiconductor material converting light to electricity. Two main types of silicon used are amorphous and crystalline. Crystalline cells are either single crystalline or multicrystalline. Within each group of technology several variants may be distinguished. The commercial silicon feedstock available to solar cell is of two types – the metallurgical grade (silicon metal) and the semiconductor grade (polysilicon). Metallurgical grade silicon contains high concentrations of intentional impurities such as boron and phosphorous and unintentional impurities of transitional metal atoms and other elements. Unintentional impurities are called the "kilier impurities" as they reduce the minority carrier life time. Impurity presences in high concentrations prevent the use of metallurgical grade silicone in solar cells.

Semiconductor grade silicon is more refined and the impurity levels are at the ppb level. Although purity requirements for solar cells are not as stringent as for semiconductors, the industry has up to now been forced to select its raw material from the semiconductor grade silicon. As the semiconductor grade silicon is very costly, and it contributes an appreciable fraction of the cost of solar cells, the photovoltaic energy conversion has become an expensive process. Silicon used for solar cells can tolerate higher concentrations of impurities than in semiconductor grade, without affecting the efficiency of the cell considerably. The silicon having these impurities in substantial amount, but within tolerable limits is called the **solar grade silicon**, which is generally viewed as a material which is still very pure, although somewhat less than semiconductor grade silicon. Therefore, solar grade silicon lies in between the metallurgical grade silicon and semiconductor grade silicon. There is more and more interest for silicon of rather low purity and low cost as a starting material for solar cells.

Lot of efforts have gone into design a specific route to solar grade silicon between the two existing commercial grades. The approach can be upgrading the purity of metallurgical silicone to the accepted levels for solar grade or altering and simplifying the process of preparing the semiconductor grade silicon to get solar grade polysilicon. The challenge in these processes is to achieve sufficient out put at acceptable cost.

Production of metallurgical grade silicon (Carbothermic reduction of silica)

Metallurgical grade silicon with a typical purity of 98.5% Si is produced in submerged electrical arc furnace. The furnace consists of a crucible filled with quartz and carbon (metallurgical coke and/or coal). Silicon is formed as per the following reaction:

$$SiO_2(s) + 2C(s) \rightarrow Si(l) + 2CO(g)$$

The raw material mix is heated by means of an intense electric arc sustained between the tip of three submerged electrodes and the electrical ground of the furnace. Electrodes are also made of carbon. Liquid silicon is tapped from the bottom of the furnace, and mixed raw materials are charged from the top.

Refining

The crude silicon liquid is taken in large ladles and treated when still liquid with silica sand (SiO_2) and lime/limestone ($CaO/CaCO_3$). Elements less noble than silicon such as Al, Ca and Mg are oxidized.

$$4Al + 3SiO_2 \rightarrow 3Si(l) + 2Al_2O_3$$

$$2Ca + SiO_2 \rightarrow Si(l) + 2CaO$$

$$2Mg + SiO_2 \rightarrow Si(l) + 2MgO$$

After completion of oxidative refining in the ladle, the slag containing the impurities is removed mechanically and liquid silicon is poured into a casting mold and solidified. The solidified silicon is crushed into small lumps up to 100 mm, in jaw crushers and roll crushers.

Production of semiconductor grade silicon

Impurities in the ppb or ppt range are required for polysilicon used in semoconductor industry. This is achieved by the preparation of a volatile silicon hydride and its purification generally using fractional distillation. This is followed by the decomposition of this hydride to hyper pure elemental silicon by reductive pyrolysis or chemical vapor deposition.

Siemens process

The most popular process is based on the thermal decomposition of trchlorosilane at 1100°C on a heated silicon rod placed inside a deposition chamber. This process is commonly referred to as the Siemens process with reference to the company that carried out its early developments.

Trichlorosilane is prepared by hydrochlorination of metallurgical grade silicon in a fluidized reactor.

$$Si(s) + 3HCl \rightarrow HSiCl_3 + H_2$$

This reaction occurs at 350°C normally without a catalyst. The trichlorosilane formed is double purified through fractional distillation. In the first step, the heavier components resulting from the direct synthesis are removed. In the second step, components lighter than trichlorosilane are eliminated.

High purity $SiHCl_3$ is then vaporized, diluted with hydrogen and introduced into the deposition reactors. The gas is decomposed onto the surface of heated silicon seed rods, electrically heated to about 1100°C, and growing large rods of hyper pure silicon.

The reactions involved are:

$$2SiHCl_3 \rightarrow SiH_2Cl_2 + SiCl_4$$

$$SiH_2Cl_2 \rightarrow Si + 2HCl$$

$$H_2 + HSiCl_3 \rightarrow Si + 3HCl$$

$$HCl + HSiCl_3 \rightarrow SiCl_4 + H_2$$

Union Carbide process

A more recent method developed by Union Carbide Chemicals in the united States of America, replaces the trichlorosilane with silane (SiH_4). But the principle of decomposition on a heated silicon rod inside a closed deposition chamber is maintained.

The process involves the following steps:

The hydrogenation of tetrachlorosilane through a bed of metallurgical silicon is carried out in a fluidized bed reactor.

$$3SiCl_4 + 2H_2 + Si \rightarrow 4SiHCl_3$$

The trichlorosilane is separated by distillation while the unreacted tetrachlorosilane is recycled back to the hydrogenation reactor.

The purified trichlorosilane is passed through a fixed bed column filled with quaternary ammonium ion exchange resin acting as catalyst. Trichlorosilane gets converted into dichlorosilane.

$$2HSiCl_3 \rightarrow H_2SiCl_2 + SiCl_4$$

The products are separated by distillation, tetrachlorosilane is recycled to the hydrogenation reactor and dichlorosilane is passed through a second fixed bed column filled with quaternary ammonium ion exchange resin. Dichlorosilane is converted into silane.

$$3H_2SiCl_2 \rightarrow SiH_4 + 2HSiCl_3$$

The above products are separated by distillation and trichlorosilane is recycled to the first fixed bed column. Silane is further purified by distillation and then pyrolized to produce polysilicon onto heated silicon seed rods mounted in a metal bell-jar reactor.

$$SiH_4 \rightarrow 2H_2 + Si$$

Purification by zone refining

Silicon can be purified by the zone refining technique. The method involves a large number of successive crystallization from the melt. This method is used when the impurities are more soluble in the melt than in the solid material. In a bar of impure silicon held vertically, a short zone at the top is melted by supplying heat from an external source. The molten zone is moved slowly down the length of the solid from the top to the bottom, when it is finally frozen. The process is repeated as many times as needed. The impurities collect preferentially in the liquid zone, move with the molten zone and accumulate at the end of the rod where they are finally allowed to freeze. Successive passes of molten zone lead to further purification, as the liquid will be less pure than the solid crystallizing from it. After the process is over, the bottom portion, concentrated with impurities is removed.

If boron is present as an impurity, zone refining method can not be used for its removal, because, its solubility is same in the molten and the solid silicon.

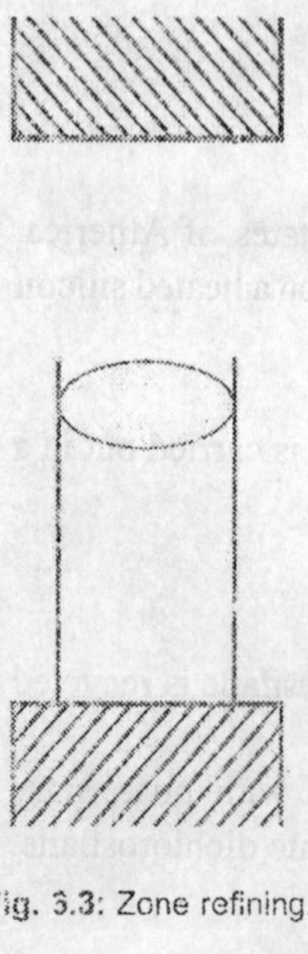

Fig. 3.3: Zone refining.

Doping of Si

There are several methods to incorporate the desirable impurities in the crystals of semiconductors so that desirable extrinsic semiconductor properties are obtained. These techniques are known as doping techniques. Doping can be carried out at the crystal formation stage itself, by adding calculated amount of dopants into the melt. When silcon is deposited on to the surface of silicon rod by vapor decomposition, doping can be accomplished by simultaneously depositing a dopant with the semiconductor material. This can be done by mixing the reaction mixture gas feed with vapors of suitable compounds of the dopant. For example, calculated amount of PH_3 is mixed with the gas feed to get n-type doping and BH_3 for p-type doping. Some of the other methods used are explained in the following sections.

Diffusion technique

In this technique, a region of a semiconductor material is incorporated with dopant atoms by the diffusion of impurity atoms into the crystal of the material without actually melting it. By this technique, the extent of impurity penetration can be controlled to a very small thickness of the material. For example, a n-type silicon can be obtained by heating a silicon wafer below its melting point in an atmosphere of n-type impurities such as phosphorous. The impurity atoms condensed on the surface of the wafer, diffuses into the crystal. Similarly p-type silicon can be obtained by heating in an atmosphere of p-type dopants. The extent of diffusion can be regulated by controlling the temperature and the concentration of the impurity atoms.

Ion implantation technique

In this technique, the semiconductor is impacted with an ion beam of impurity ions. This results in the implantation of some dopant atoms into the semiconductor crystal. The extent of implantation is controlled by the energy of the ion beam.

REVIEW QUESTIONS

1. Differentiate between direct solar power utilization and indirect solar power utilization
2. What are the advantages and disadvantages of solar power?
3. What is a photovoltaic cell? Explain its working.
4. Discuss the importance of the photovoltaic cell.
5. What are the advantages and disadvantages of a photovoltaic cell?
6. What are solar cells, modules, panels and arrays?
7. Write a note on physical and chemical properties of silicon relevant to photovoltaics.
8. Explain the following: (i) Metallurgical grade silicon (ii) Semiconductor grade silicon and (iii) Solar grade silicon.
9. Explain the method for the production of metallurgical grade silicon.
10. Explain Siemens process and Union carbide process of preparing semiconductor grade silicon.
11. Explain the methods of doping silicon.

CHAPTER 7

High Polymers

7.1. INTRODUCTION

Man has passed through many ages such as stone age, bronze age, steel age and at present he is living in a most fascinating world of polymers. The days are not far off when plastics are more likely to replace the conventional materials. There is no exaggeration if one calls the present era as the era of plastics.

High polymers are natural and synthetic macromolecules which play a vital role in our every day life. These wonder materials provide the basic needs of our lives, namely, food, clothing and shelter. They possess wide range of properties from soft like silk to strong like steel. Because of this wide spectrum of polymer properties, one can make a variety of articles from simple toys to tough tyres. There is a class of plastics called thermoplastics which can be shaped by application of heat and pressure without affecting their properties. Many plastics come in attractive colors of rainbow with all possible shades and designs. The fabricated articles have imported look and feel. Further, they have long useful life and are priced economically.

Of late, the markets are flooded with synthetic high polymers which are not the gift of nature but creations of chemists. The chemist with his magic tools of test tubes and flasks created new high polymers in his laboratory and passed the secrets to chemical technologist who in turn produced them on commercial scale. In the last sixty years or so, no industry has progressed as much as polymer industry.

The art of polymerization is of recent origin. The new techniques developed not only helped to imitate natural polymers but also to produce a variety of new polymers which were ever made by nature. The first modified natural polymer, cellulose nitrate was produced commercially around 1860. The first synthetic polymer '**Bakelite**' entered the market around 1910.

The major systematic development in the polymer science and technology took place since 1920 and this lead to the production of plastics, fibers, elastomers, biopolymers. The latest addition to the ever expanding field is the production of conducting polymers having conductivity as high as that of copper metal.

7.2. SOME BASIC DEFINITIONS

Polymers

A polymer is a giant or macro molecule formed by the repeated union of several simple molecules (monomers). The repeat units in a polymer chain are linked through strong covalent bonds. The properties of polymers are entirely different from those of its monomers. Average molecular mass of a commercial polymer is normally in the range of $10^3 - 10^7$. Polyethylene, polystyrene, teflon, plexiglass, etc., belong to the family of polymers.

Monomers

A monomer is a simple molecule having two or more bonding sites through which each can link to the other monomers to form a polymer chain. Monomers are often called **"building blocks"** of a polymer chain. All simple molecules, however, do not qualify as monomers. The essential requirement of a small molecule to qualify as a monomer is the possession of at least two bonding sites. In other words, a monomer can undergo polymerization only if it is at least bifunctional. Vinyl chloride, styrene, adipic acid, hexamethylene diamine belong to the family of monomers because these possess two active centers in the form of double bonds or functional groups.

Polymerization

The fundamental chemical process by which the monomers (low molecular weight) are converted into polymers (high molecular weight) is called the polymerisation reaction. All polymerization reactions need suitable initiators.

The terms monomer, polymer and polymerization reaction are illustrated below:

$$\underset{\text{Ethylene (Monomer)}}{nCH_2 = CH_2} \xrightarrow[\text{Initiator}]{\text{Polymerization}} \underset{\text{Polyethylene (Polymer)}}{[-CH_2 - CH_2 -]_n}$$

Functionality

The total number of functional groups or bonding sites present in a monomer molecule is called the functionality of the monomer. Depending on the number of bonding sites, the monomers are called bifunctional, trifunctional or polyfunctional. A simple molecule should possess at least bifunctionality, which is the basic requirement to qualify as a monomer and to undergo polymerization.

The double bonds ($C = C$) in all vinyl compounds provide two bonding sites. Therefore all vinyl compounds are bifunctional monomers. The functionality of adipic acid ($HOOC - [CH_2]_4 - COOH$)

Table 7.1. Functionality of Some Commercial Monomers

Monomer	*Functional group Reactive Centers*	*Functionality*
1. Vinyl Compounds	Two reactive centers at the double bond	Two
2. Ethylene glycol	Two alcoholic OH groups	Two
3. Glycerol	Three alcoholic OH groups	Three
4. Adipic acid	Two-COOH groups	Two
5. Hexamethylene diamine	Two-NH_2 groups	Two
6. Phenol	Three active centers at ortho & para positions	Three
7. Formaldehyde	Two active centers	Two
8. 2,4,6 – Trimethylol phenol	Three alcoholic OH groups	Three
9. Bisphenol A	Two Phenolic – OH	Two
10. Diisocyanate	Two isocyanate groups	Two

and hexamethylene diamine ($H_2N - [CH_2]_6 - NH_2$) is two because each contains two reactive functional groups. The functionality of some important monomers are listed in Table 9.1.

Degree of Polymerization

The degree of polymerization is a number which expresses the total number of repeat units in a polymer chain. It is denoted by DP.

$$n[CH_2 = CH_2] \rightarrow [-CH_2 - CH_2 -]n$$
$$\text{Ethylene} \qquad \text{Polyethylene}$$

Where 'n' is the degree of polymerization.

The value of DP indicates the length of a polymer chain. The DP can also be used to compute the molecular weight of a polymer. The molecular weight of a polymer is the product of the number of repeat units (DP) times the molecular weight of repeat unit.

$$\text{Molecular weight of polymer} = \text{DP} \times \text{mol.wt. of repeat unit.}$$

If the DP of polystyrene is 1,000, its molecular weight is calculated as under:

$$\text{Mol. wt. of polystyrene} = \text{DP} \times \text{Mol. wt. of Styrene}$$
$$= 1000 \times 104 = 1{,}04{,}000$$

7.3. CLASSIFICATION OF POLYMERS

Polymers are complex macro molecules. They have different properties in contrast to simple compounds from which they are formed. Majority of modern day polymers are organic polymers.

These polymers are classified in different ways based on their sources, thermal behaviour, mechanism of polymerization and properties as under:

(a) Natural and synthetic polymers
(b) Thermoplastic and thermosetting polymers
(c) Addition and condensation polymers
(d) Elastomers, fibers, resins and plastics

(a) Natural and Synthetic Polymers: This type of classification is based on the source of polymers. The polymers which are obtained from natural sources such as plants and animals are called natural polymers.

Examples: The polymer materials such as cotton, wool, silk, starch, cellulose, proteins, natural rubber, leather, etc., belong to the family of natural polymers.

The polymers which are synthesized from simple molecules are called synthetic polymers. Majority of modern day commercial polymers are synthetic polymers.

Examples: Polyethylene, Polystyrene, Teflon, PVC, Plexiglass, Nylons, Polyesters, etc., belong to synthetic polymers family.

(b) Thermoplastics and Thermosettings: The basis for this classification is the thermal behaviour of the polymers.

Thermoplastics: The polymers which become soft on heating and hard on cooling are called thermoplastics. This cycle can be carried out many times without affecting their chemical properties.

Examples: Polyethylene, plexiglass, polyvinyl chloride, teflon, etc., are thermoplastics.

Thermosettings: The polymers which undergo chemical changes and cross linking on heating and become permanently hard, rigid and infusible are called thermosettings. They will not soften on heating, once they are set.

Examples: Phenol-formaldehyde resin, urea-formaldehyde resin, epoxy resins, etc., are typical examples of thermosettings.

(c) Addition and condensation polymers: The basis for this classification is their polymerization mechanism.

The polymers which are formed by the addition polymerization reaction in which self addition of several olefinic monomers to each other takes place without elimination of by-products are called addition polymers.

Examples: The polymers such as PVC, Polystyrene, Plexiglass, Teflon, Polyethylene, etc., belong to addition polymers family.

The polymers which are formed by intermolecular condensation reaction through functional groups of monomers with continuous elimination of by-products are called condensation polymers.

Examples: Nylons, Polyesters, Phenol-formaldehyde resin, urea – formaldehyde resin, melamine formaldehyde resin, etc., are common examples for condensation polymers.

(d) Elastomers, Fibers, Plastics, and Resins: Polymers can also be classified based on their structure and end uses.

(i) Elastomers: The polymer which undergo a very large elongation when pulled, but return to the original length on release of force is called an elastomer. The polymer chains of elastomers are long, coiled and entangled. There are no intermolecular forces except weak van der waals forces.

Examples: Natural rubber, Buna-S, Neoprene rubber, Nitrile rubber, Silicone rubber, etc.

(ii) Fibers: Fibers are long, thin and thread like polymer chains which do not undergo stretching or deformation like elastomers. These long chain molecules are lined up and held together by hydrogen bonding. Fiber-forming materials may be synthetic or natural ones.

Examples: Wood, silk, cotton, jute, etc., are the examples of natural fiber forming materials. Nylon 66, Nylon 6, Terylene, Orlox, ete., are synthetic fiber forming materials.

(iii) Plastics: The polymers which can be molded into desired articles by the application of heat and pressure are called plastics. Plexiglass, Polyvinyl chloride, Teflon, etc., are familiar examples of plastics.

(iv) Resins: Low molecular mass polymers either in the liquid or solid form are called resins. Resins are used as adhesives or molding powders.

Examples: P-F resin, U-F resin, Epoxy resin, M-F resin.

7.4. TYPES OF POLYMERIZATION

The monomers can undergo polymerization by two types, with each one having a different and distinct mechanism. They are:

(a) Addition polymerization
(b) Condensation polymerization.

Of late, the terms addition and condensation polymerization are modified as chain polymerization and step polymerization.

(a) Addition (chain) polymerization

A polymerization reaction in which rapid self addition of several bifunctional monomers to each other takes place by chain reaction without the elimination of any by products is called addition or chain polymerization. The monomers containing double bonds (vinyl compounds) are the one which undergo addition polymerization.

$$n\,[CH_2 = \underset{\displaystyle y}{\underset{|}{CH}}] \xrightarrow{\text{Addition polymerization}} [-CH_2 - \underset{\displaystyle y}{\underset{|}{CH}} -]_n$$

Vinyl compound → Vinyl polymer

where Y = H, Ethylene
CH_3, Propylene
Cl, Vinyl Chloride
C_6H_5, Styrene
CN, Acrylonitrile

The main features of chain polymerization are:

- Only olefin compounds can undergo addition polymerization.
- The double bond provides required two bonding sites. There is absolutely no elimination of any by products.
- The self addition of monomers takes place very rapidly.
- The addition polymerization is brought about by free radicals, ionic or co-ordination mechanism.
- Linear polymers are produced.
- The polymer produced has the same elemental composition as that of the monomer.
- The molecular weight of polymer is an integral multiple of that of the monomer.

Copolymerization

In addition polymerization if only one type of monomer is used, the polymer formed contains the same repeat units in the polymer chain. Such polymers are called homopolymers. The polymers such as PVC, PTFE, PE, etc., are homopolymers. On the other hand, it is also possible to polymerize two different monomers to produce a new polymer called copolymer. They are free radical chain polymerization reactions.

The process of simultaneous polymerization of two or more different types of monomers (olefinic) is called copolymerization and the polymer which contains different units is called a copolymer. The properties of a copolymer are different from those of individual homopolymers.

$$n\,CH_2 = \underset{\displaystyle Cl}{\underset{|}{CH}} + m\,CH_2 = \underset{\displaystyle OCOCH_3}{\underset{|}{CH}} \xrightarrow{\text{Copolymerization}} [CH_2 - \underset{\displaystyle Cl}{\underset{|}{CH}}]_n - [CH_2 - \underset{\displaystyle OCOCH_3}{\underset{|}{CH}}]_m$$

VC   VAC   Poly (vinylchloride-vinyl acctate) copolymer

The widely used synthetic rubbers, SBR, Nitrile rubber are other examples of copolymers. The copolymers may be (i) Alternating (ii) Random (iii) Block and (iv) Graft Copolymers.

(i) Alternating Copolymer: A polymer in which the two different repeat units arranged alternatively throughout the chain is called an alternate copolymer. It has an orderly arrangement and linear structure.

$$M_1 - M_2 - M_1 - M_2 - M_1 - M_2 - M_1 - M_2 -$$

where M_1 and M_2 are vinyl chloride(VC) and vinyl acetate [VAc) monomers respectively.

(ii) Random Copolymer: A polymer in which the two repeat units are arranged in random fashion throughout the chain, is called random copolymer. Random copolymers have no orderly arrangement and are with linear structure.

$$M_1 - M_1 - M_1 - M_1 - M_2 - M_2 - M_1 - M_1 - M_2 - M_2 - M_2 -$$

where, M_1 and M_2 are styrene and 1, 3-butadiene respectively.

(iii) Block Copolymer: A polymer in which a block of one repeat unit is followed by a block of another repeat unit consecutively is called a block copolymer.

$$M_1 - M_1 - M_1 - M_1 - M_2 - M_2 - M_2 - M_2 - M_1 - M_1 - M_1 - M_1 - M_2 - M_2 - M_2 - M_2 -$$

where M_1 and M_2 are acrylic acid and methyl methacrylate respectively.

Block copolymers have orderly arrangement and have linear structure.

(iv) Graft Copolymer: A polymer in which the main polymer chain consists of only one type of repeat units while the branch is made up of yet another repeat unit is called a graft copolymer.

$$\begin{array}{ccccccccccccc} & & & & & (M_2)_n & & & (M_2)_n & & & & \\ & & & & & | & & & | & & & & \\ - & M_1 & - & M_1 & - & M_1 & - & M_1 & - & M_1 & - & M_1 & - \\ & | & & & & & & & & & & & \\ & (M_2)_n & & & & & & & & & & & \end{array}$$

M_1 = VC
M_2 = styrene

where styrene repeat units are grafted on polyvinyl chloride.

Mechanism of Free-radical Polymerization

The complete polymerization of olefinic monomers by free radical mechanism proceeds in three distinct stages:

(a) Initiation (b) Propagation (c) Termination.

(a) Initiation: The polymerization is initiated by free radicals generated from initiators such as peroxides. A free radical is an atomic or molecular species having an odd or unpaired electron. They are highly active species. An initiator on heating undergoes homolytic cleavage and generates free-radicals. The initiator free radical adds to a vinyl monomer to produce a new free-radical.

$$\underset{\text{Initiator}}{R - R} \xrightarrow{\Delta} \underset{\text{Free radical}}{2R^{\bullet}}$$

$$\underset{\text{Free radical}}{R^{\bullet}} + \underset{\text{vinyl monomer}}{CH_2 = \underset{\underset{y}{|}}{CH}} \longrightarrow \underset{\text{New free radical}}{R - CH_2 - \underset{\underset{y}{|}}{CH}}$$

(b) Propagation: In the propagation step, the new free-radical attacks another monomer molecule to produce yet another new free radical. Many repetitions of this step lead to polymer chain growth and continue until termination occurs.

$$R - CH_2 - \underset{\underset{y}{|}}{\overset{\bullet}{CH}} + CH_2 = \underset{\underset{y}{|}}{CH} \longrightarrow R - CH_2 - \underset{\underset{y}{|}}{CH} - CH_2 - \underset{\underset{y}{|}}{CH} + CH_2 = \underset{\underset{y}{|}}{CH}$$

$$\longrightarrow R - CH_2 - \underset{\underset{y}{|}}{CH} - CH_2 - \underset{\underset{y}{|}}{CH} - CH_2 - \underset{\underset{y}{|}}{CH}$$

The structure of the growing chain can be represented as

$$R - (CH_2 - \underset{\underset{y}{|}}{CH} -)_n \, CH_2 - \underset{\underset{y}{|}}{\overset{\bullet}{CH}} \text{ or } R \ldots CH_2 - \underset{\underset{y}{|}}{\overset{\bullet}{CH}}$$

(c) Termination: The last-stage in free radical polymerization is the termination step. In the termination process the growing polymer chain is transformed into a dead polymer molecule. The growth of the polymer chain can be arrested (chain termination) by one of the following processes.

(i) Coupling of two growing chains: Termination of macro free-radical chains usually takes place by coupling of two growing chain radicals.

$$R\ldots CH_2-\underset{y}{\underset{|}{\dot{C}H}}+\underset{y}{\underset{|}{\dot{C}H}}-CH_2\ldots R \longrightarrow R\ldots CH_2-\underset{y}{\underset{|}{CH}}-\underset{y}{\underset{|}{CH}}-CH_2\ldots R$$

Growing Chains — Dead Polymer

(ii) Coupling of growing chain with initiator free radical

$$R\ldots CH_2-\underset{y}{\underset{|}{\dot{C}H}}+\dot{R} \longrightarrow R\ldots CH_2-\underset{y}{\underset{|}{CH}}-R$$

Growing Chains — Initiator radical — Dead Polymer

(iii) Disproportionation

$$R\ldots CH_2-\underset{y}{\underset{|}{\dot{C}H}}+\underset{y}{\underset{|}{\dot{C}H}}-CH_2\ldots R \longrightarrow R\ldots CH=\underset{y}{\underset{|}{CH}}+\underset{y}{\underset{|}{CH_2}}-CH_2\ldots R$$

The termination of free-radical chain polymerization may also take place by disproportionation. In this termination process, a hydrogen atom from one growing chain is transferred to another growing chain, resulting in one of the 'dead' polymers having an unsaturated end.

(iv) Coupling of growing chain with inhibitors: Inhibitors are chemical substances capable of inhibiting the chain growth by combining with the growing chains and forming stable products or dead polymer. Examples: hydroquinone, nitrobenzene, dinitrobenzene, benzothiazine, etc.

Atmospheric oxygen can act as a good inhibitor and can inhibit the polymerization reaction. Therefore, free radical polymerization is carried out in the absence of oxygen, under a nitrogen atmosphere. The inhibiting action of oxygen is attributed to its bi-radical nature. If $G^{\bullet}$ is a growing chain, then,

$$G^{\bullet} + O^{\bullet} - O^{\bullet} \rightarrow G - O - O^{\bullet}$$

$$G - O - O^{\bullet} + G^{\bullet} \rightarrow G - O - O - G$$

Mechanism of ionic polymerization

The ionic mechanism of addition polymerization involves the attack on the π electron pair of the monomer by either a positive ion (cation) or a negative ion (anion). Correspondingly, the mechanisms are known as cationic mechanism and anionic mechanism respectively.

Cationic mechanism

In cationic mechanism strong Lewis acids such as BF_3 act as the initiator or is generally called the catalyst. The mechanism can be given in three steps as initiation, propagation and termination.

(a) Initiation: The Lewis acid BF_3 in the presence of small amounts of water (known as co-catalyst) forms hydrate which exists as an ion pair.

$$\underset{\displaystyle F}{\overset{\displaystyle F}{F-\underset{|}{\overset{|}{B}}}} + H-O-H \longrightarrow \left[F-\underset{\underset{\displaystyle F}{|}}{\overset{\overset{\displaystyle F}{|}}{B}}-OH \right]^{-} \text{----}\ H^{+} \text{ OR } [FBOH]^{-} \text{ ---- } H^{+}$$

where H^+ is the initiating cationic species. The H^+ ion combines with the monomer to form the carbonium ion.

$$[F_3BOH]^{-} \text{-----}\ H + CH_2 = CH_2 = \underset{\underset{\displaystyle X}{|}}{CH} \longrightarrow CH_3 - \underset{\underset{\displaystyle X}{|}}{CH} + [F_3\,BOH]$$

(b) Propagation: The propagation involves further adding up of the monomer units and simultaneous transfer of the charge to the newly added monomer unit.

$$CH_3 - \underset{\underset{\displaystyle X}{|}}{\overset{+}{CH}} + [F_3BOH]^{-} + CH_2 = \underset{\underset{\displaystyle X}{|}}{CH} \longrightarrow CH_3 - \underset{\underset{\displaystyle X}{|}}{CH} - CH_2 - \underset{\underset{\displaystyle X}{|}}{\overset{+}{CH}} + [F_3BOH]^{-}$$

As more and more monomer units are added up, the chain keeps on growing and the growing chain can be represented as follows:

$$R \text{ --- } CH_2 - \underset{|}{CH}$$

As the polymer chain grows, the π electron pair is shifted in a direction opposite to the growth of the chain, C^+ ion keeps on moving in the direction of chain growth and the counter ion, $[F_3BOH]^-$, moves along with the C^+ ion.

(c) Termination: Termination occurs when collision occurs between the growing carbonium ion and an anion takes place. The possible termination reactions are:

(i) By the donation of a proton to the counter ion (ion pair precipitation): The growing chain donates a proton to the counter ion, resulting in the formation of a double bond at the end of the polymer molecule, arresting the growth of the chain.

$$\underset{\text{Growing chain}}{K \text{ ---- } CH_2 - \underset{\underset{\displaystyle X}{|}}{\overset{+}{CH}} + [F_3BOH]^{-}} \longrightarrow \underset{\text{Dead polymer molecule}}{\underset{\underset{\displaystyle X}{|}}{R} \text{ ----- } CH = CH + F_3\,BOH]^{-} \text{ ----- } H^{+}}$$

The process of donation of a proton and the reformation of the BF_3 hydrate is called "ion pair precipitation".

(ii) By simple coupling: This type of termination reaction involves the formation of a covalent bond between the carbonium ion and the counter ion.

$$R \text{-----} CH_2 - \overset{+}{\underset{X}{CH}} + [F_3BOH]^- \longrightarrow R \text{-----} CH_2 - \underset{X}{\overset{OH}{CH}} + BF_3$$

The initiator is regenerated in both types of termination. In addition to BF_3, compounds like $AlCl_3$, $SnCl_4$ and $TiCl_4$ can also be used as initiators. Examples of monomers undergoing cationic polymerization include isobutylene, styrene, methyl styrene and many vinyl ethers.

Anionic mechanism

In anionic mechanism, organo alkali compound (alkyl and aryl derivatives of alkali metals) such as butyl lithium (LiBu) is used as initiators or the catalyst. The mechanism can be given in three steps as initiation, propagation and termination.

(a) Initiation: The organo alkyl compound, LiBu, ionizes to liberate the butyl anion.

$$LiBu \rightarrow Li^+ + Bu^-$$

where Bu^- is the initiating anionic species. The Bu^- ion combines with the monomer to form the carbanion.

$$Bu^- + CH_2 = \underset{X}{CH} \longrightarrow Bu - CH_2 - \underset{X}{\overset{-}{C}H}$$

(b) Propagation: The propagation involves further adding up of the monomer units and simultaneous transfer of the negative charge to the newly added monomer unit.

$$Bu\text{-}CH_2 - \underset{X}{\overset{-}{C}H} + CH_2 = \underset{X}{CH} \longrightarrow Bu - \underset{X}{CH} - CH_2 - \underset{X}{\overset{-}{C}H}$$

As more and more monomer units are added up, the chain keeps on growing and the growing chain can be represented as follows:

$$Bu - [CH_2 - \underset{X}{CH}]_n - CH_2 - \underset{X}{\overset{-}{C}H}$$

As the polymer chain grows, the π electron pair is shifted in the same direction as that of the chain growth, C^- ion keeps on moving in the direction of chain growth and the counter ion, Li^+, moves along with the C^- ion.

(c) Termination: The termination in anionic polymerization is not a spontaneous process. Unless some impurities are present or some strongly ionic substances are added deliberately, no termination occurs. If the polymerization is carried out under controlled conditions and impurities are avoided, the reaction proceeds till all the monomers are consumed and the carbanions at the chain ends remain potentially active. If a fresh quantity of monomer is added the polymerization again continues till the freshly added monomers are consumed. Therefore, the polymers formed by this technique are called "**living polymers**". They are 'alive' and keep on growing as long as monomers are supplied.

In addition to butyl lithium, triphenyl methyl potassium and ethyl sodium are also used as initiators. Monomers capable of undergoing anionic polymerization include butadiene, isoprene, styrene and acrylonitrile.

(b) Condensation (step) Polymerization

A polymerization reaction in which bi- or poly functional monomers undergo inter molecular condensation with continuous elimination of by products (H_2O, HCl, NH_3, C_2H_5OH, CH_3OH) is called condensation or step polymerization. They are normally catalyzed by acids or alkali. Since condensation reaction is rather slow and proceeds stepwise, such polymerization is often called step polymerization. The monomers such as ethylene glycol, terphthalic acid, hexamethylene diamine, adipic acid, etc., which have two required functional groups can be polymerized by step polymerization process.

Typical examples for condensation polymerization are:

(i) $$n\,H_2N-(CH_2)_6-NH_2 + n\,HOOC-(CH_2)_4-COOH$$

Hexamethylene diamine — Adipicacid

$$[-HN-(CH_2)_6-NH-CO-(CH_2)_4-CO-]_n + 2n\,H_2O$$

Nylon 66

(ii) $$n\,HO-(CH_2)_2-OH + n\,HOOC-C_6H_4-COOH$$

Ethyleneglycol — Terphthalic acid

$$[-O-(CH_2)_2-OOC-C_6H_4-CO-]_n + 2n\,H_2O$$

Polyester

The main features of condensation polymerization are:

- The monomers having two or more reactive functional groups can undergo condensation polymerization.
- Polymerization proceeds through intermolecular condensation.
- There is continuous elimination of by products.
- The polymer chain build up is slow and step wise.
- Polymerization is catalyzed by acids or alkali.
- Linear or cross linked polymers are produced.
- The elemental composition of the polymer is generally different from that of its monomers.

7.5. METHODS OF POLYMERIZATION

Industrial polymerization can be carried out by different techniques. The principal polymerization methods are:

(a) Bulk polymerization
(b) Solution polymerization
(c) Suspension polymerization
(d) Emulsion polymerization.

The choice of a particular method depends on the nature of monomer, the polymer formed and the polymerization mechanism followed. Each method has its inherent advantages and disadvantages.

(a) Bulk polymerization

In this method, polymerization occurs in homogeneous system. The method is used for the polymerization of liquid monomers. The reaction mixture consists of a liquid monomer which acts as solvent for the added initiator and chain transfer agent if any used. The entire reaction mass is kept under constant agitation for effective heat and mass transfer. It is heated initially to polymerisation temperature to initiate the reaction. The polymer product formed, being free from any contamination except the monomer, can be used without purification.

Advantages

- The method is relatively simple and requires simple equipment.
- The conversion is extremely high (= 100%).
- The purity of the product is high.
- The product needs neither isolation nor purification.
- The product has high optical clarity.

Disadvantages

- The agitation is difficult
- The heat control is more difficult. As the reaction progresses, the viscosity increases and hot spots tend to develop due to the difficulty in agitation. The uncontrolled exothermic reactions may lead to explosion.
- The gel effect is observed. As the polymerization progresses, the viscosity increases, the termination becomes difficult and rate of polymerization increases rapidly. This phenomenon of auto acceleration in polymerization rate with decrease in termination rate is called "**gel effect**" or "**auto acceleration**".
- A broad molecular weight distribution in polymer occurs.
- This method is practiced for relatively less exothermic polymerization reactions.

Applications: The bulk polymerization is mainly used for the manufacture of condensation polymers where the reactions are mildly exothermic. The method can also be used for the production of plexiglass and polystyrene.

(b) Solution polymerization

In solution polymerization, like bulk polymerization, the polymerization occurs in homogeneous system. The presence of solvent facilitates the heat and mass transfer and reduces viscosity build up.

The monomer, free radical initiator and chain transfer agent, if any used; are dissolved in a good inert solvent. The reaction mass is kept under constant agitation and heated to a temperature of polymerization to initiate the reaction. After the completion of the reaction, the product is isolated from the solvent by evaporation or used as such in solution form.

Advantages

- The viscosity build up is negligible.
- The heat control is easy.
- The agitation is easy.
- The product in solution can be directly used as adhesives and coatings.

Disadvantages

- The product isolation needs evaporation of solvent.
- Requires solvent recovery and recycling.
- The product is contaminated with solvent since the removal of the last traces of solvent is often difficult.
- It is difficult to produce a very high molecular weight polymer since the solvent may act as a chain transfer agent, thereby limiting the molecular weight.

Applications: This method is widely practised for the production of polyethylene, polyvinyl chloride, poly acrylonitrile and block co-polymers. The solutions of some polymers produced are directly used as adhesives, coatings and acrylic ester finishes.

(c) Suspension (Pearl) polymerization

In this method, unlike the bulk and solution polymerization, polymerization occurs in heterogeneous system and the method is used only for water insoluble monomers. Since the product separates out as spherical pearls or beads, the method is often called "pearl polymerization". The use of aqueous media facilitates agitation and heat control.

The water insoluble monomer is suspended in water as several tiny droplets by continuous agitation. These droplets are prevented from coalescing by the use of small amounts of water soluble polymers such as polyvinyl alcohol or colloids. The initiators used are soluble in monomers. Each monomer droplet containing dissolved initiator, being discrete, the method in effect is bulk polymerization. The reaction mass is heated to initiate the polymerization. The product formed being water insoluble separates out in the form of beautiful spherical '**pearls**' or '**beads**'. They are isolated by the easy method of filtration, followed by washing.

Advantages

- The viscosity build up of polymer is negligible.
- Efficient thermal control.
- Isolation of product is easy as it needs only filtration and washing.
- Higher purity product.
- The process is more economical since water is used in place of costly solvents.
- Isolated products need no further purification.

Disadvantages

- The method is applicable only for polymerization of water insoluble monomers.
- It is difficult to control particle size.

Applications: The method is used for commercial production of polyvinyl chloride, polyvinyl acetate, styrene-divinyl benzene (ion exchanger), etc.

(d) Emulsion Polymerization

This is an important method for the production of a large number of commercial plastics and elastomers. The process offers the unique opportunity to produce a polymer of high molecular weight. The essential ingredients are monomers (water insoluble), initiators (water soluble) such as potassium persulfate, H_2O_2, etc., and surface active agents – soaps and detergents (RCOONa, RSO_3Na). The water insoluble monomer is dispersed in water not as discrete droplets but as an uniform emulsion. The emulsion is stabilized by adding surface active agents. After adding the initiator, the system is kept agitated in the absence of oxygen at 70°C (Fig. 7.1).

Mechanism: The surfactant is normally a soap or a detergent. It has hydrophilic head and hydrophobic tail. At very low concentration, it gets completely dissolved in water. At a little higher concentration it gets dispersed. When the concentration of surfactant exceeds a critical micelle concentration (CMC), the soap molecules form micelles (aggregation of 50–100 molecules) oriented with tails inwards and heads outward. The interior of a micelle is provided with hydrocarbon environment.

When the monomer is added, bulk of it remains in the aqueous phase but some of it dissolves in each of the micelle. The free radicals are generated in the aqueous phase.

$$H{-}O{-}O{-}H \rightarrow 2OH^{\bullet}$$

$$S_2O_8^{2-} + Fe^{2+} + \rightarrow Fe^{3+} + SO_4^{2-} + SO_4^{\bullet-}$$

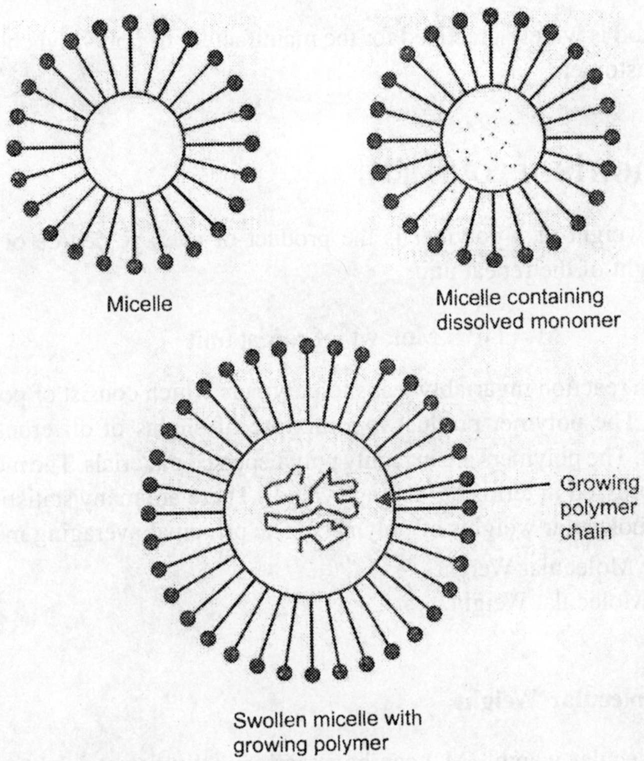

Fig. 7.1. Emulsion polymerization

The radicals diffuse through the aqueous phase, penetrate into the micelle and initiate the polymerization in each micelle. The polymer growth inside the micelle causes its swelling. The monomers consumed inside the micelle are replenished by continuous diffusion of fresh monomers from the aqueous phase. The polymer growth continues until the growing chain is encountered by a radical and gets terminated. The polymer formed is in the form of well stabilized latex. It is isolated either by coagulation using electrolytes or by freezing.

Advantages

- Easy agitation.
- Easy heat control.
- The rate of polymerization is high
- A very high molecular weight polymer is obtained due to the continuous supply of monomers.
- Molecular weight control is possible.

Disadvantages

- The polymer needs additional clean up and purification.
- It is difficult to remove entrapped coagulants, emulsifiers, etc.
- The method often requires rapid agitation.

Applications: The method is widely practiced for the manufacture of polyvinyl chloride, polyvinyl acetate and synthetic elastomers.

7.6. MOLECULAR WEIGHTS OF POLYMERS

The average molecular weight of a polymer is the product of average degree of polymerization times the molecular weight of the repeat unit.

$$M = DP \times \text{Mol. wt. of repeat unit}$$

However, polymerization reaction invariably leads to polymers which consist of polymer chains of widely varying lengths. The polymer product is a mixture of chains of different lengths but of definite molecular weight. The polymers are in reality polydispersed materials. The molecular weights of such polymers are expressed in terms of 'average value'. There are many statistical methods for calculating the average molecular weights of polymers. The principal averaging methods are:

(a) Number Average Molecular Weight.
(b) Weight Average Molecular Weight.

(a) Number Average Molecular Weight

The number average molecular weight (M_n) can be experimentally determined by using different techniques such as osmometry, freezing point lowering, boiling point elevation and end group analysis.

M_n is calculated by dividing the sum of the molecular weight contribution from the individual molecules by the number of molecules. The details of calculations are as shown below.

Let a polymer chain consist of 'n' polymer chains of which n_1 of M_1 mol. wt., n_2 of M_2 mol. wt...... till n_i of M_i molecular weight.

Total number of molecules $= n_1 + n_2 + n_3 + \ldots\ldots\ldots + n_i = \Sigma n_i$

Number fraction of n_1 species $= \dfrac{n_1}{\sum n_i}$

Molecuar wt. contribution from n_1 species $= \dfrac{n_1 M_1}{\sum n_i}$

The molecular weight contribution from other species, computed in similar fashion, yields

$$\frac{n_2 M_2}{\sum n_i}, \frac{n_3 M_3}{\sum n_i}, \ldots \frac{n_i M_i}{\sum n_i}$$

Therefore,

$$M_n = \frac{n_1M_1}{\sum n_i} + \frac{n_2M_2}{\sum n_i} + \dots + \frac{n_iM_i}{\sum n_i}$$

$$= \frac{\sum n_iM_i}{\sum n_i}$$

(b) Weight Average Molecular Weight

Weight average molecular weight is determined by experiments in which each molecule or chain makes contribution to the measured results relative to its size. The average is more dependent on the number of heavier molecules than is the number average molecular weight which is simply dependent on total number of molecules.

Weight of n_1 species = $W_1 = n_1M_1$

Similarly the weights of n_2, n_3, ...n_i species are, n_2M_2, n_3M_3, ... n_iM_i, respectively.

Total weight of the polymer = $W = \Sigma\, n_iM_i$

$$\text{Weight fraction of } n_1 \text{ species} = \frac{n_1M_1}{\sum n_iM_i}$$

$$\text{Molecular weight contribution from } n_1 \text{ species} = \frac{W_1M_1}{W} = \frac{n_1M_1 \times M_1}{\sum n_iM_i} \quad \frac{n_1M_1^2}{\sum n_iM_i}$$

Similarly mol. wt. contributions from other species are

$$\frac{n_2M_2^2}{\sum n_iM_i}, \frac{n_3M_3^2}{\sum n_iM_i}, \dots\dots\dots, \frac{n_iM_i^2}{\sum n_iM_i}$$

The weight average molecular weight is given by the expression,

$$M_w = \frac{n_1M_1^2}{\sum n_iM_i} + \frac{n_2M_2^2}{\sum n_iM_i} + \dots\dots\dots + \frac{n_iM_i^2}{\sum n_iM_i} = \frac{\sum n_iM_i^2}{\sum n_iM_i}$$

Problem

A polymer sample contains 1,2,3 and 4 molecules having molecular weights 10^5, 2×10^5, 3×10^5 and 4×10^5, respectively. Calculate the number average and weight average molecular weights of the polymer.

Solution:

$$M_n = \frac{\sum n_i M_i}{\sum n_i}$$

$$= \frac{(1 \times 10^5) + 2 \times (2 \times 10^5) + 3 \times (3 \times 10^5) + 4 \times (4 \times 10^5)}{1 + 2 + 3 + 4}$$

$$= \frac{(1 + 4 + 9 + 16) \times 10^5}{10} = \frac{30 \times 10^5}{10} = 3.0 \times 10^5$$

$$M_w = \frac{\sum n_i M_i^2}{\sum n_i M_i}$$

$$= \frac{(1 \times 10^5)^2 + 2(2 \times 10^5)^2 + 3(3 \times 10^5)^2 + 4(4 \times 10^5)^2}{(1 \times 10^5) + 2(2 \times 10^5) + 3(3 \times 10^5) + 4(4 \times 10^5)}$$

$$= \frac{(1 + 8 + 27 + 64) \times 10^{10}}{(1 + 4 + 9 + 16)} = \frac{100 \times 10^{10}}{30 \times 10^5}$$

$$= \mathbf{3.3 \times 10^5}$$

7.7. GLASS TRANSITION TEMPERATURE

All amorphous polymers, when cooled below a certain temperature, become stiff, hard, brittle and glassy but above this temperature, they are soft, flexible and rubbery. This transition temperature of a polymer is called **'glass transition'** temperature. It is denoted by T_g. The hard, brittle state is known as the glassy state and the soft flexible state as the rubbery state. In the glassy state, there are neither segmental nor molecular motions and virtually all chain motions are completely frozen.

On heating beyond T_g, the polymer passes from the glassy state into the rubbery state. In rubbery state, there are only segmental motions while molecular mobility is forbidden. On further heating much above T_g, each polymer chain eventually obtains sufficient energy, molecular mobility sets in and the melt polymer starts flowing. The temperature below which it is in rubbery state and above which it behaves like a viscous liquid is called the **melting point** of the polymer, denoted by T_m. Since no polymer has sharp melting point, the transition temperature at which polymer passes from rubbery state to liquid state is called its **flow temperature** (T_f).

The plots showing change of specific volume (v) of a polymer with temperature (T) are shown in Fig. 7.2.

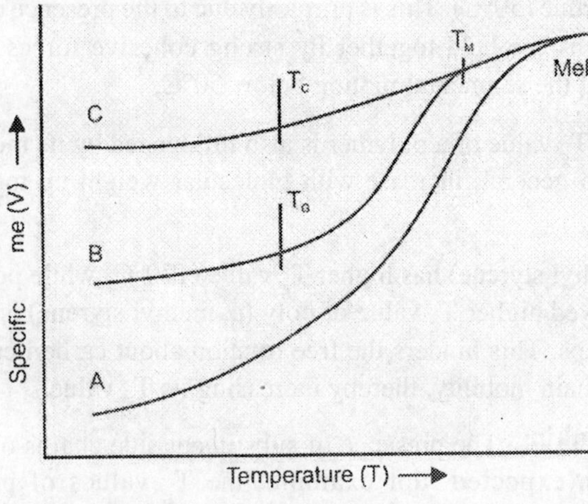

A – Crystalline;
B – Partially crystalline;
C – Amorphous polymer

Fig. 7.2. Schematic representation of variation of specific volume with temperature

Parameters Influencing T_g Value

There are several structural parameters of a polymer which influence the T_g value.

(i) Crystallinity: The T_g value of a polymer largely depends on the degree of crystallinity. Higher the crystallinity, larger is the T_g value of a polymer. Segmental and chain mobilities are easier in non-crystalline or amorphous polymers than in crystalline polymers. In crystalline polymer, the linear or stereo regular chains are lined up parallel to each other and are held by strong cohesive forces. This leads to a high T_g value of the polymer.

(ii) Flexibility: The degree of freedom with which different segments of the polymer along the chain back bone can rotate around the covalent bond is the measure of flexibility of the polymer. The higher the freedom to rotate, the more flexible are the chain segments, higher their segmental mobility, and hence, lower is the T_g value. Linear polymers made of C-C, C-N or C-O single bonds with higher degree of freedom for rotation, have lower T_g values. The presence of aromatic or cyclic structure in the chain backbone hinders the free rotation of segments, increasing the T_g value. The higher T_g value of cellulose nitrate is because of the rigid ring structures in the polymer chain of the polymer.

(iii) Intermolecular forces: The intermolecular forces determine the strength with which polymer chains are held together in a polymer sample. In hydrocarbon polymers, with only van der Waals forces holding the polymer chains together lacks cohesive forces. The chain segments can slip past each other easily. For example, polyethylene has a low T_g value (–125°C). Polymer chains containing polar groups are held together more strongly by neighbouring dipoles as well as intermolecular hydrogen bonding, and their movement is restricted. Nylon 6 having similar structure as that of polyethylene, has higher T_g value (50°C). This is primarily due to the presence of large polar groups in polymer chains. The chains are held together by strong cohesive forces through inter-chain hydrogen bonding, restricting the segmental motion below 50°C.

(iv) Molecular weight: The T_g value of a polymer is also influenced by its molecular weight. The T_g values of all polymers, in general, increase with molecular weight up to 20,000 and beyond which the effect is negligible.

(v) Side group: Poly (α-methyl styrene) has higher T_g value (170°C) while polystyrene has lower T_g value (100°C). The observed higher T_g value of poly (α-methyl styrene) is due to the presence of effective methyl side groups. This hinders the free rotation about carbon-carbon bond of chain backbone, and restricts the chain mobility, thereby increasing its T_g value.

(vi) Branching and cross linking: The presence of substituent side chains or branching exerts a different effect on Tg than expected. For example, the T_g values of polymethylacrylate, polyethylacrylate and polybutylacrylate are 10°C, –24°C and –54°C, respectively. If the hindrance for the free rotation of the polymer chain is taken into account, polymehylacrylate should have had lowest and polybutylacrylate should have had highest T_g amongst the three polymers in accordance with their bulkiness. But the trend is reverse, because, the side chains themselves are flexible and as the side chain length increases from methyl to butyl, their flexibility also increases. Also as the side chain length increases, they push the neigbhouring main chains further apart, thereby increasing the free volume and, hence, flexibility.

Cross linking of chains decreases the flexibility of the polymer chains and, therefore, as the extent of cross linking increases, the T_g value decreases.

(vii) Plasticizers: Plasticizers are low molecular weight non volatile substances, which when added to a polymer improves its flexibility and processibility. Therefore, the presence of plasticizers reduces the T_g of polymers. Plasticizer molecules penetrate into the polymer matrix, occupying the space between the polymer molecules and thereby reducing the cohesive forces between them. Without plasticizers, many of the commercially useful polymers such as PVC and nylon would have been useless as their T_g values are above ambient temperature and are brittle at normal temperature.

(viii) Stereo-regularity: Stereo-regular polymers can be arranged more orderly and better close packed, compared to the stereo-irregular polymers. Therefore stereo-regular polymers have higher crystallinity, and hence will have higher glass transition temperature.

Importance of T_g

The T_g value is an important parameter of polymers. The polymers above their T_g value are soft and flexible while below T_g, they are hard and brittle. Thus T_g value is a measure of flexibility of polymers. The use of any polymer at any temperature is decided by its T_g value. T_g value along with Tm helps in choosing the right temperature of processing the polymers by different techniques.

7.8 STRUCTURE-PROPERTY RELATIONSHIP

The structure of a polymer has profound influence on some of the properties of polymers. The properties such as crystallinity, tensile strength, impact resistance, plasticity and chemical resistance are largely dependent on the structure of the polymer and also on its molecular weight. This structure-property relationship of polymer is briefly discussed in the following sections.

(a) Crystallinity

Polymers invariably contain both crystalline and amorphous regions. The degree of crystallinity of a polymer depends on its structure (linear, branched, with large pendant groups in polymer chain) and configuration (stereo regular or not). Crystalline regions occur when linear polymer chains without branching and carrying no bulky groups, are orderly arranged parallel and close to each other. The chains of polymer are held together by secondary forces such as van der waals forces, hydrogen bonding, polar interaction, etc. Such type of close packing imparts a high degree of crystallinity. The polymers having high degree of crystallinity exhibit high tensile strength, impact resistance, high density and sharp and high melting point (T_m).

The polymers such as HDPE, stereo regular isotactic and syndiotactic isomers of poly propylene, PVC are highly crystalline (Fig. 7.3).

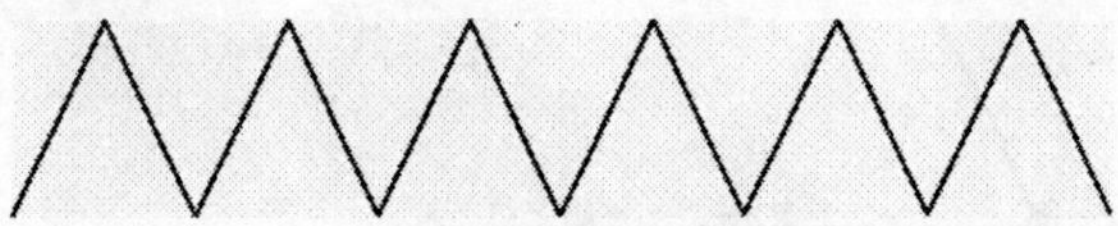

(a) HDPE

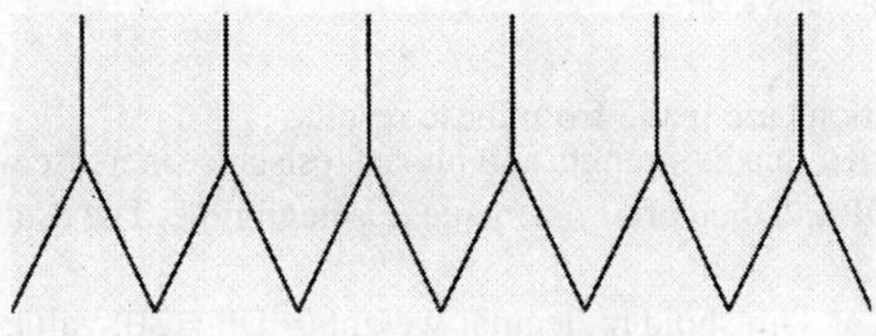

(b) Isotactic polypropylene

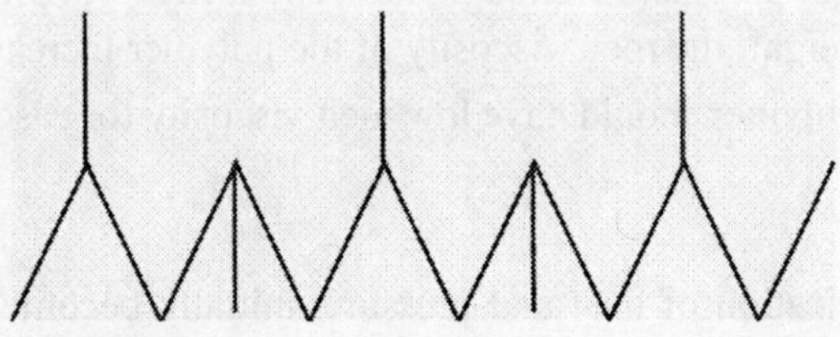

(c) Syndiotactic

Fig. 7.3. Linear, stereoregular crystalline polymers

Atactic PVC, PS, polypropylene in which bulky pendant groups arranged randomly on the polymer backbone are amorphous. LDP which has extensive branching is also amorphous. The randomly arranged bulky groups and branches on polymer matrix, prevent close packing of chains, resulting in amorphous regions. Inter chain forces are weak in these polymers.

(b) Tensile strength

The tensile strength is one of the important properties of useful commercial polymers. This property is directly dependent on the molecular weight (DP) of the polymer. The relationship of this property with molecular weight of the polymer is shown in the plot (Fig. 7.4).

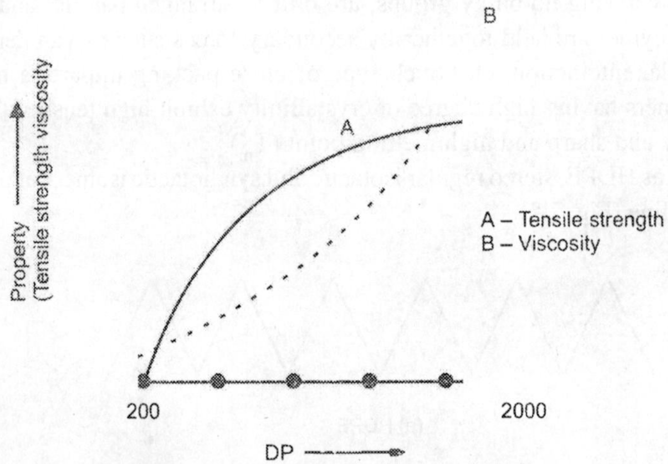

Fig. 7.4. Plot of mechanical properties Vs DP

The following observations are made from these results:

- Mechanical properties, tensile strength and impact resistance increase with molecular weight (DP) up to ≈ 2000DP and thereafter the increase is negligible. The useful range is 200–2000 DP.
- Every polymer has a 'Threshold molecular weight (≈ DP = 30) value below which it does not possess useful strength.
- High molecular weight polymers are tough, hard and difficult to process.
- At high molecular weight, the melt viscosity of the polymer increases steeply.

A commercially useful polymer should have low melt viscosity for ease of processing.

(c) Plastic Deformation

Some polymers, on the application of heat and pressure, initially become soft, flexible, rubbery matter and undergo deformation. On further heating beyond T_m, they melt and flow. Such a property of the polymer is called plasticity. On cooling, they return to their original state. This property of plastic deformation is used in molding operations. Thermoplastics exhibit this property. Thermoplastics are linear, stereo regular polymers. The polymer chains are closely packed and held by secondary forces such as van der waals, hydrogen bonding and dipolar interaction. Such polymers when heated, the chains acquire sufficient energy and overcome these inter chain attractive forces. They attain molecular mobility and flow like viscous fluid.

Thermosettings do not exhibit plasticity. This can be explained on the nature of their structure. Molded thermosettings have three dimensional structure. All monomer units are held together through strong primary covalent bonds throughout the structure. Primary covalent bonds are not easily broken by heat. On strong heating, charring occurs instead of deformation. Therefore thermosettings do not undergo reversible plastic deformation.

(d) Chemical Resistivity

The chemical resistance of a polymer depends on the composition and structure of the polymer and also on the nature of the attacking reagents. The hydrocarbon polymers having no polar groups are generally swell or dissolve in hydrocarbon solvents like gasoline, benzene, cyclohexane, etc. (like dissolves the like). The polymers containing polar groups are attacked by polar solvents. Starch, polyvinyl alcohol, polyvinyl acetate, etc., are soluble in polar solvents.

Polymers containing residual unsaturation undergo oxidative degradation when exposed to air and light. The air (O_2 / O_3) and UV light attack the residual double bonds in the polymer chains and cause oxidative degradation of polymers. Natural rubber and synthetic rubbers which contain residual unsaturation are susceptible for oxidative degradation in service conditions.

Denser packing makes the penetration of solvent or chemical substance more difficult into the polymer material. Therefore more crystalline the polymer, higher is the chemical resistance. Higher the degree of cross linking in the polymer, higher is its chemical resistance.

(e) Elasticity

Elastic character of a polymer is the measure of its ability to recover the original shape after a deformation stress is released. The elastomers are characterized by the high degree of elasticity. The elastic deformation in polymers arises from the typical coiled and entangled structure of the polymer chains. In the normal unstretched state, the polymer chains are in a random arrangement. When stretched, the coiled chains begin to disentangle and straighten out. As a result, a better orientation of the chains takes place resulting in crystallization, which in turn increases the force of attraction between different chains, thereby causing stiffening of the material. When the strain is released, the stretched chains return to their original coiled and entangled form.

7.9 RESINS AND PLASTICS

(a) Resins

Resins are linear, low molecular weight, soluble and fusible polymers. They are often called prepolymers. These prepolymers during curing undergo further polymerization and extensive cross linking, resulting in the formation of an hard and rigid three dimensional network. The resins are used as adhesives, coatings and molding powders.

Typical examples of resins are:

Amino resins (UF, MF), phenolic resins (PF), epoxy resins and thermosetting polyester resins used in paints.

(b) Plastics

Plastics are high polymers which can be molded by the application of heat and pressure. There is no chemical change during the molding process. They have linear structures.

Examples of plastics are: Polyethylene, Polystyrene, Polyvinyl Chloride, Nylons, Polyesters, etc.

Thermoplastics: Thermoplastics are the polymers which become soft on heating and hard on cooling and the process can be repeated for a number of times. They undergo reversible changes on heating.

Examples of Thermoplastics: Teflon, Plexiglass, Polystyrene, PVC, Nylons, Polyesters.

Thermosetting resins: Thermosetting resins are polymers which are fusible on initial heating, but on further heating become permanently set insoluble, infusible, hard, rigid and three dimensional network product. They undergo irreversible changes on heating.

Examples of thermosetting resins are: Urea – Formaldehyde resin (UF), Melamine – Formaldehyde resin (MF), Phenol – Formaldehyde (PF) resin, Epoxy resin, thermosetting polyester resins, etc. The other differences of thermoplastics and thermosettings are given in Table 7.2

Table 7.2. Differences between thermoplastics and thermosettings

	Thermoplastics	*Thermosettings*
1.	They soften on heating and harden on cooling	They are fusible on initial heating, but become permanently hard, rigid and infusible product on further heating
2.	They undergo reversible changes on the application of heat.	They undergo irreversible changes on the application of heat.
3.	They can be reshaped by heat cycle.	They can not be reshaped by heat cycle.
4.	They are soft and flexible.	They are hard, rigid and infusible.
5.	They have linear structure.	They have three dimensional net work.
6.	There is no change in chemical composition and structure during molding process.	They undergo chemical changes such as further polymerization and cross linking during molding process.
7.	They swell or dissolve in organic solvents.	They neither dissolve nor swell in organic solvents.
8.	The molded articles are to be cooled to room temperature before taking out from the moulds to avoid deformation.	The molded articles can be taken out of the moulds even when they are still hot without any deformations.
	Examples Polyethylene, Polypropylene, Polystyrene, Polyvinyl chloride, Teflon, Plexiglass	*Examples* Phenol – Formaldehyde resin (PF) Urea – Formaldeyde resin (UF) Melamine – Formaldehyde resin (MF), Epoxy resins.

7.10. SOME COMMERCIAL POLYMERS

(a) Polyethylene (PE)

Polyethylene is the simplest homo polymer and a thermoplastic. It is represented by the following structure.

$$(-CH_2-CH_2-)_n$$

There are two commercial grades of polyethylene. They are available as **'low density'** polyethylene (LDPE) and **'high density'** polyethylene (HDPE). They differ in a number of properties.

(i) Low density polyethylene (LDPE): LDPE is manufactured by polymerization of ethylene under high pressure (1500 – 50,000 psi) and high temperature (350°C) using a trace of oxygen or any peroxide such as benzoyl peroxide as initiator.

$$\underset{\text{Ethylene}}{n\,CH_2=CH_2} \xrightarrow[\text{trace } O_2]{\text{Chain polymerization}} \underset{\text{LDPE}}{[-CH_2-CH_2-]_n}$$

Properties: LDPE has a linear structure with extensive branching. It has a low degree of crystallinity (40%), low density (0.912–0.935) and low melting point (110°C). Further, it has lower tensile strength, stiffness and hardness. Its chemical resistance is also poor.

(ii) High density polyethylene (HDPE): HDPE is manufactured from ethylene by coordination polymerization using Ziegler Natta catalyst [$TiCl_4 + Al(C_2H_5)_3$ in hydrocarbon solvent].

Table 7.3. Distinguishing properties of LDPE and HDPE

	LDPE	*HDPE*
(i)	Produced by chain polymerization using O_2 or peroxides.	Produced by coordination polymerization using Ziegler-Natta catalyst.
(ii)	Linear polymer with extensive branching.	Linear polymer with practically no branching.
(iii)	Relatively lower degree of crystallinity (40%)	Relatively higher degree of crystallinity (90%)
(iv)	Lower melting point (110°C)	Higher melting point (140°C)
(v)	Lower density (0.912 – 0.935)	Higher density (0.96)
(vi)	Lower tensile strength, stiffness and hardness.	Higher tensile strength, stiffness and hardness.
(vii)	Chemical resistance is low	Chemical resistance is high.

$$n\,CH_2 = CH_2 \xrightarrow[TiCl_4 + Al(C_2H_5)_3]{\text{Coordition polymerization}} [-CH_2 - CH_2 -]_n$$

Ethylene — HDPE

HDPE is essentially a linear polymer with practically no branching. HDPE has relatively a high degree of crystallinity (90%), high density (0.96) and high melting point (140°C).

It also has high tensile strength, stiffness and hardness. It is not attacked by chemicals.

Applications

(i) **Films and sheets:** A large amount of polyethylene produced is used as packing materials for food and textile.

(ii) **Moulded articles**: It is also used for the manufacture of moulded articles such as bottles, cans, buckets, toys, tubs, pipes, squeeze bottles, etc.

(iii) **Insulating material**: Because of its excellent electrical insulation properties, it is used for wire and cable coating.

(b) Polytetrafluoroethylene (PTFE) (Teflon)

'Teflon' is the trade name for polytetrafluoroethylene. Teflon has the following linear structure.

$$[-CF_2 - CF_2 -]_n$$

The monomer used is tetrafluoro ethylene which is a non-toxic gas (b.p –76°C). The Monomer (TFE) is obtained by the following reactions:

$$CHCl_3 + 2HF \rightarrow CHClF_2 + 2HCl$$

Chloroform — Chlorodifluoro methane

$$2CHClF_2 \xrightarrow{\text{Pyrolysis}} CF_2 = CF_2 + 2HCl$$

Tetrafluoroethylene

Teflon is usually manufactured by emulsion polymerization of tetrafluoroethylene using peroxide as initiators:

$$n\,CF_2 - CF_2 \xrightarrow[\text{Peroxide}]{\text{Emulsion polymerization}} [-CF_2 - CF_2 -]_n$$

TFE — Teflon

Properties: Teflon is essentially a linear polymer with no branching. It is a thermoplastic. In contrast to other polymers, teflon has the following unique properties:

- High degree of crystallinity (93–98%).

- High melting point (T_m 327°C).
- High density (2.30 g/cm^3).
- Practically insoluble in any solvent.
- High resistance to corrosive chemicals, oxidizing and reducing agents and strong mineral acids including fuming nitric acid.
- Not wetted by either oil or water.
- Remains slippery over a wide range of temperature (– 40 to 300°C).
- Low coefficient of friction.
- High thermal stability, good mechanical strength.
- Excellent electrical insulating properties.

Teflon is, therefore, rightly called a **'Wonder plastic'**.

Teflon has impracticably very low melt flow rates above T_m due to a very high viscosity. It is, therefore, fabricated by an unconventional method.

Since teflon has cold flow, it is fabricated by pressing the polymer in cold into the desired shape, followed by sintering at temperature above T_m (380°C) to yield dense, strong and homogeneous sample.

Applications

(i) Insulating material: It is used for insulation of motors, generators, transformers, coils, capacitors, wires and cables.

(ii) Antistic applications: Teflon is not wetted by oil or water. Teflon is coated on articles such as bakery trays, frying pans and food-processing equipments.

(iii) Industrial filters, gaskets and belts of teflon are used, where resistance to chemical and high temperature are required.

(iv) Anticorrosive coating: Teflon can be sprayed on metal surfaces just like paints and then heat fused into a coating that sticks like enamel. These coatings offer permanent protection against corrosion. This technique is widely used for army weapons.

(v) Lubricant: Teflon is an ideal lubricant because it stays slippery over a wide range of temperature (– 40 to 300°C). It does not collect dust and sand. It is therefore used for non lubricated bearings and as a dry lubricant.

(c) Poly (methyl methacrylate)

Plexiglass is the trade name for poly (methyl methacrylate). It has the following linear structure.

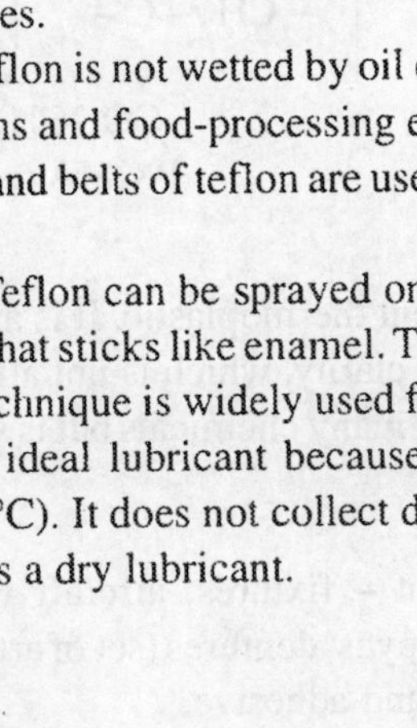

Monomer, methyl methacrylate is obtained by the following reactions:

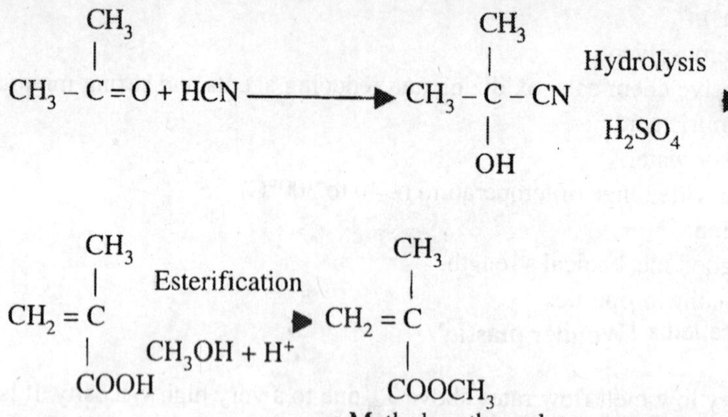

Methyl methacrylate

Polymerization: PMMA is manufactured by bulk or suspension polymerization methods using peroxides as initiators.

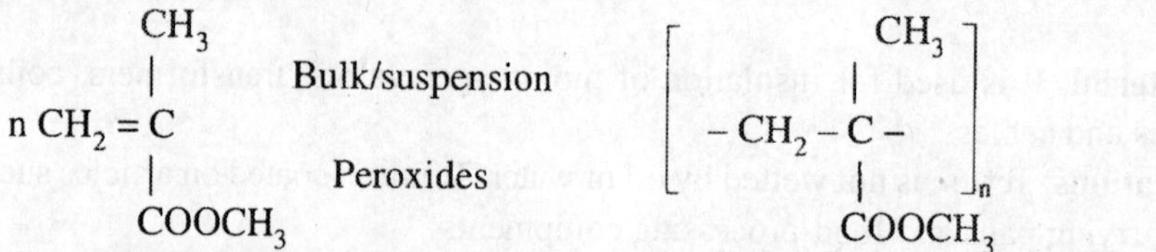

Methyl methacrylate PMMA (plexi glass)

Properties: Plexiglass is a white transparent thermoplastic. It is amorphous in nature due to bulky pendant groups. It has an excellent optical clarity, which is not affected by sunlight or age but has poor scratch resistance. It is not attacked by many chemicals but is soluble in many organic solvents. It has a good dimensional stability as well.

Applications: It is used for making light – fixtures, aircraft windows, attractive signboards, instruments and appliance covers, artificial eyes, dentures (set of artificial teeth), transparent moulds and tubes. Plexiglass is also used as paint and adhesive.

(d) Polyurethanes

Polyurethanes belong to an important and most versatile family of polymers. They received attention because of their diverse applications. Polyurethanes are characterized by the presence of **'Urethane'** linkage (–NH–CO–O–) in the molecular chains. They have linear structure and are represented by the following structure.

$$[-\underset{\underset{O}{\|}}{C}-NH-(-R-)-NH-\underset{\underset{O}{\|}}{C}-O-(-R'-)-O-]_n$$

They are prepared by the poly addition reaction between **diisocyanate** and a **diol** or a **triol.** During the addition, the 'H' atom of 'OH' group migrates to and adds to nitrogen atom as shown in the following scheme.

$$\underset{\text{di-isocyanate}}{O=C=N(-R-)N=C=O} + \underset{\text{diol}}{HO-(R')-OH}$$

$$[-\underset{\underset{O}{\|}}{C}-NH-(-R-)-NH-\underset{\underset{O}{\|}}{C}-O-(-R'-)-O-]_n$$

Polyurethane

The polymerization is an addition reaction without elimination. The urethane linkage (– NH – CO – O –) resembles closely with that of polyamide linkage (– NH – CO –). However the presence of extra oxygen in urethane linkage offers increased flexibility to the chain. This is the cause for its much lower melting point than polyamides.

Applications

(i) **Elastomers:** Because of their resistance to oil, grease and corrosive chemicals, they are used for tyre treads and industrial wheels.

(ii) **Fibers:** Perol-U is a typical polyurethane fiber used in light weight garments and swim suits because of its stretching property.

(iii) **Coatings:** It is used as floor coating for gymnasium and dance floors where high abrasion resistance is required.

(iv) **Foams:** Polyurethane foams are widely used as cushions for furniture and automobiles.

(e) Polycarbonates

Polycarbonates are polyesters of phenols and carbonic acid. They can be prepared by the condensation of bisphenol-A with diphenylcarbonate or dimethyl carbonate as follows:

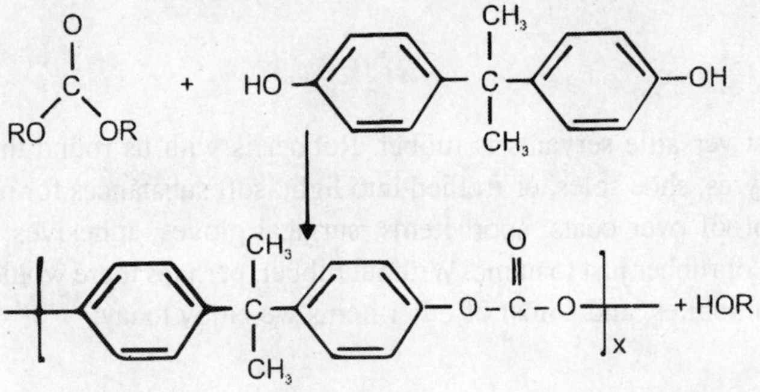

Polycarbonates are also prepared by the reaction of sodium salt of bisphenol-A with phosgene as per the following reaction:

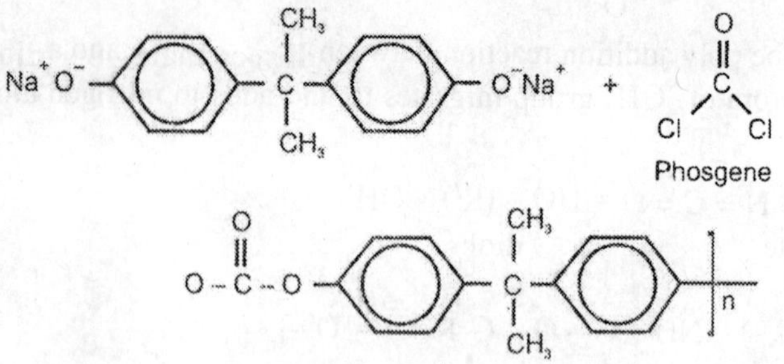

Properties:

- Polycarbonates maintain good mechanical properties between -40 °F and 280 °F.
- They possess high strength that make them resistant to impact and fracture.
- They are resistant to water and many organic solvents.
- They can be easily coloured.
- They are non-toxic, and can be absolutely transparent up to 2 inch in width.
- They also feature high electrical and heat resistance.
- They are biologically inert.
- They are readily recyciable.

Applications:

- Polycarbonate's outstanding strength makes it suitable for bullet-resistant or shatter-resistant glass applications.
- Their relatively low weight in comparison to other high strength materials and its high ductility makes it attractive to be used in lenses and windows.
- They have also been used as a flame retardant and an electrical insulators.
- They are used in small filters for the extrusion of small particles.
- They are used in CDs, DVDs, automotives, cell phones, and laptop parts.

7.11. ELASTOMERS (RUBBERS)

Introduction

One of the man's most versatile servants is rubber. Rubber is with us round the clock. It can be made hard and tough tyres, shoe soles, or frothed into light, soft substances for mattresses, pillows and cushions. Water proof over coats, sport items, surgical gloves, adhesives, combs, electrical fittings are a few items of rubber just to name. Without rubber, perhaps there would be no television, telephone, radios, automobiles, and a host of other items we enjoy today.

The rubber was known to man since early 19th century. The people of some south east nations used to play with bouncing balls made from gum produced by rubber trees, which found to be water proof as well. The Portuguese people tried to make water proof hats by coating them with rubber stuff. But unfortunately sun heat melted the rubber and ruined their beautiful garments.

Despite the curious interest the rubber aroused, very little work was done to stop the rubber stuff becoming soft and sticky in summer and hard and brittle during winter. However chance discovery of vulcanization by Samuel Goodyear in 1839 suddenly unfolded a whole range of possibilities and research on rubber.

Definition

Elastomers are high polymers which undergo very large elongation (500–1000%) under stress, yet regain original dimension (size) fully on release of stress.

This property of elastomers is called elasticity. This arises due to the coiled structure of elastomers.

Characteristics of Elastomers

- They have linear but highly coiled structure.
- Being hydrocarbons, inter chain cohesive forces are negligible
- They have high degree of elasticity.
- They can retract rapidly.

Natural Rubbers

Natural rubber is found in several species of rubber trees grown in tropical countries. Of these "Hevea brasiliensis" is the most important source of natural rubber and the rubber obtained from this is often called "Hevea" rubber. Rubber is stored in the form of white fluid called 'latex' behind the bark of the rubber tree. The trees are called weeping trees as latex oozes out when a cut is made on the bark.

Natural rubber is a high polymer of isoprene (2 – methyl – 1,3 – butadiene). It is a hydrocarbon polymer having molecular formula $(C_5H_8)_n$. The DP of natural rubber is about 5000 and it has a broad distribution of molecular weight. Since poly isoprene contains C=C bonds, it exists in two geometric isomeric structures, cis and trans forms. Natural rubber (NR) is soft and has cis-configuration while 'Guttapercha' or 'Ballata' has trans configuration (Fig. 7.5).

Deficiencies of Natural Rubber

The virgin natural rubber is as useless as 24 carat gold. As such it has virtually no practical utility value. It is mainly because of its many inherent deficiencies, which are listed below.

- It is hard and brittle at low temperature, and soft and sticky at high temperature.
- It has poor keeping qualities as it develops obnoxious odor on storage.

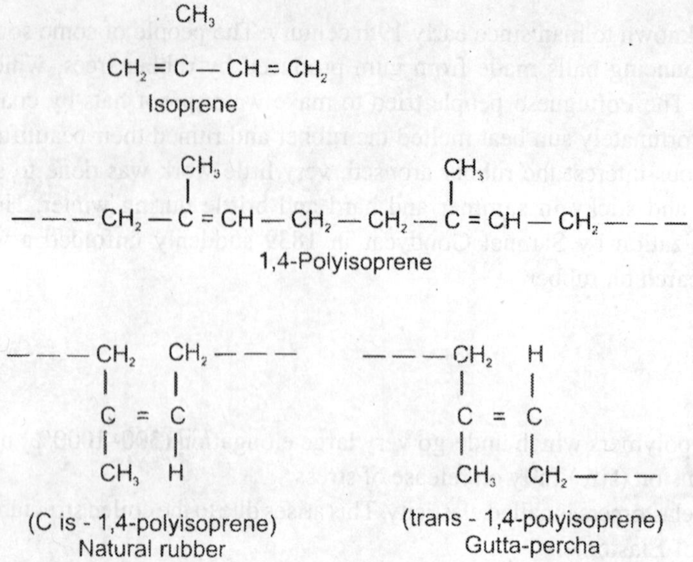

Fig. 7.5. Structures of isoprene, 1,4 – polyisoprene, cis and trans 1, 4 – polyisoprenes

- It is soft and soluble in many hydrocarbon solvents and carbon disulfide.
- It has high water absorption power.
- It is a sticky substance, rather difficult to handle.
- It is readily attacked by acids and alkalies.
- It has low tensile strength.
- It has poor abrasion resistance.
- It is sensitive to oxidative degradation.

Vulcanization

Because of the many deficiencies of natural rubber, pure rubber has practically no applications. However, the utility value of natural rubber could be immensely improved if vulcanized.

Rubber vulcanization is a process of heating the raw rubber, mixed with sulfur and other agents at 120–150°C for 1– 4 hours. It is believed that sulfur cross links (bridges) are introduced between the polymer molecules of rubber, which involve residual double bonds of polymer chains (Fig. 7.6).

The degree of vulcanization introduced in vulcanized rubber depends on the amount of sulfur used. Tyre industry uses 3–5% sulfur. If the amount of sulfur is increased to 30%, a hard and rigid rubber called 'ebonite' is produced.

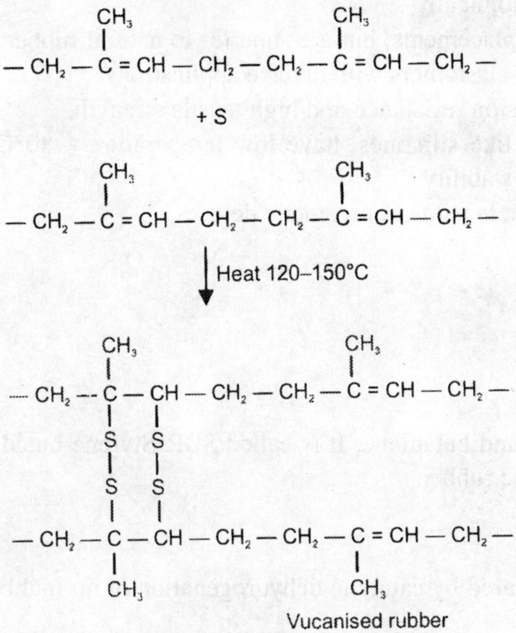

Fig. 7.6. Sulfur vulcanization of rubber.

7.12. SYNTHETIC RUBBERS

One of the greatest chemical engineering achievements of all time, second only to atomic bomb, is the discovery of synthetic (man made) rubbers. Until the world war second, natural rubber faced no competition and South East Asia was the major supplier of natural rubber to many countries. Around 1939, suddenly the supply of natural rubber was cut off to western countries. These countries felt the need to develop a technology to produce a new rubber that can replace natural rubber. The efforts of scientists and technologists resulted in the production of first ever synthetic rubber which came to the rescue of these countries.

The first useful synthetic rubber was Buna –S (GRS or SBR). It is a copolymer of 1,3 – butadiene and styrene. It was found to be an all purpose rubber. Another synthetic rubber, Neoprene was discovered, which was found to be closest to or even better than natural rubber. The continued search for new synthetic rubbers resulted in the production of a host of tailor made elastomers.

Advantages of Synthetic Rubbers

Because of better performance properties of synthetic elastomers, natural rubber failed to give stiff competition. Some of the advantages of synthetic rubbers are:

- They are produced from petrochemical raw materials in abundant amounts.

- They are priced economically
- They are not only replacements, but are superior to natural rubber in certain cases.
- They are tailor made elastomers with diverse applications
- They have high abrasion resistance and high tensile strength.
- Certain elastomers, like silicones, have low temperature (–80°C) flexibility and high temperature (250°C) stability.
- Silicones are valuable in surgical prosthetic devices.

Some synthetic rubbers

(a) Buna-S (SBR, GRS)

It is a copolymer of styrene and butadiene. It is called SBR-Styrene-butadiene rubber and GRS-Government controlled styrene rubber.

Monomers

(i) 1,3-butadiene is prepared by catalytic dehydrogenation of normal butane.

$$\underset{\text{n-butane}}{CH_3-CH_2-CH_2-CH_3} \xrightarrow[Cr_2O_3\,+\,Al_2O_3,\ 600°C]{\text{Dehydrogenation}} \underset{\text{1,3 –butadiene}}{CH_2=CH-CH=CH_2} + 2H_2$$

(ii) Styrene is produced by alkylation of benzene

$$\underset{\text{Benzene}}{C_6H_6} + \underset{\text{Ethylene}}{CH_2=CH_2} \xrightarrow{AlCl_3} \underset{\text{Ethyl benzene}}{C_6H_5-CH_2-CH_3}$$

$$\underset{\text{Ethyl benzene}}{C_6H_5-CH_2-CH_3} \xrightarrow[Fe_2O_3/MgO,\ 600°C]{\text{Dehydrogenation}} \underset{\text{Styrene}}{C_6H_5-CH=CH_2} + H_2$$

Polymerization: SBR is manufactured by emulsion polymerization. The monomers, butadiene (75 parts) and styrene (25 parts) are emulsified in water using soap or detergent. The reaction is initiated by peroxide initiators. Polymerization is carried out at 5°C and therefore the product is known as cold SBR.

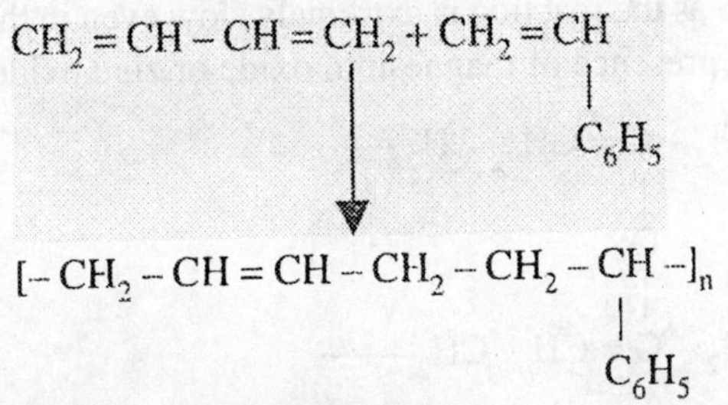

SBR is a random copolymer. It is vulcanized with sulfur. It is slightly inferior to natural rubber in its physical properties.

It is used for the manufacture of passenger car tyres and treads, motor cycle and scooter tyres, cycle tyres and tubes. They are not used for truck tyres.

They are also used for the manufacture of conveyor belts, foot-wares, shoe soles, hoses and electrical insulation.

(b) Neoprene rubber

Polychloroprene (neoprene) is a polymer of chloroprene (2-chloro-1,3-butadiene) and has the following structure

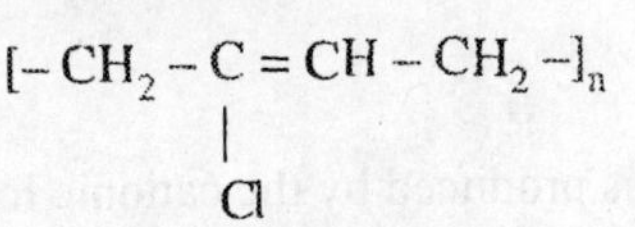

Chloroprene is produced by the catalytic addition of hydrogen chloride to vinylacetylene, which in turn is made by the catalytic dimerisation of acetylene.

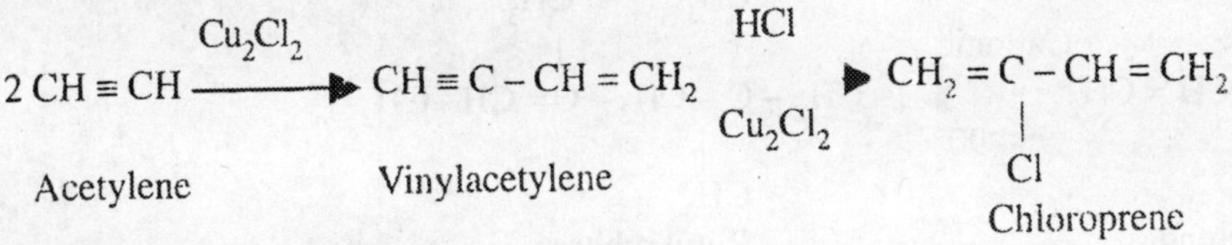

Polymerization: Polychloroprene is produced by the emulsion polymerization of chloroprene following free radical mechanism.

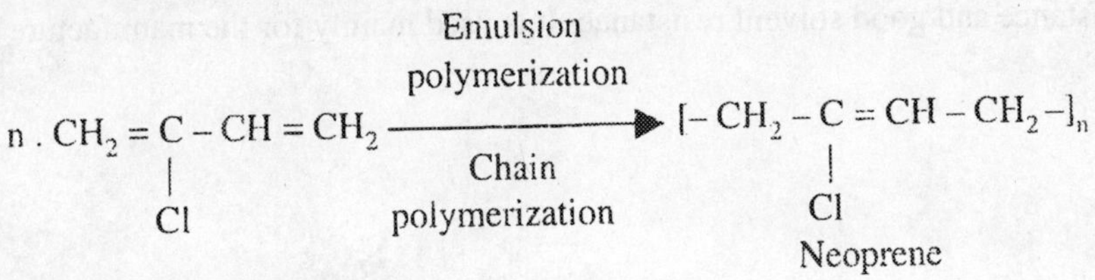

It is not vulcanized by normal sulfur vulcanization as the reaction is extremely slow even in the presence of catalysts. However it is vulcanized in the presence of magnesium oxide or zinc oxide.

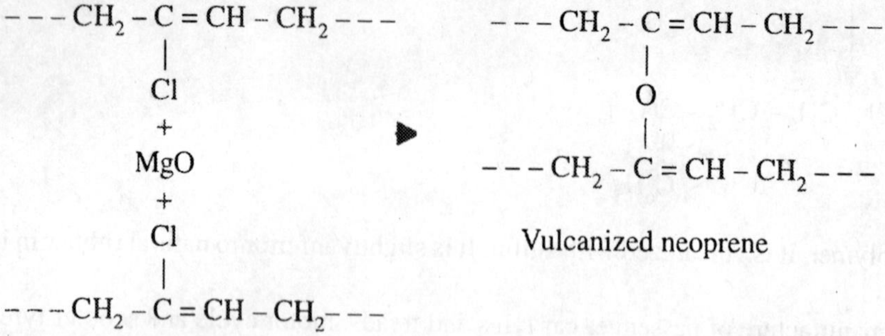

Neoprene rubber has outstanding oil and chemical resistance, high tensile strength, outstanding resistance to oxidative degradation and aging, and good ozone and weathering resistance.

Applications: Neoprene rubber can replace natural rubber in most of the applications. It is used for hoses, tubes for carrying oils and chemicals, gloves, coated fabrics, cable and wire coatings, belts, shoe heels, solid tyres.

(c) Butyl rubber

It is a copolymer of isobutylene and a small amount of isoprene. It is produced by the cationic low temperature (–90°C) copolymerization of isobutylene (90%) in the presence of a small amount of isoprene (10%). The use of isoprene is to provide necessary residual unsaturation for vulcanization.

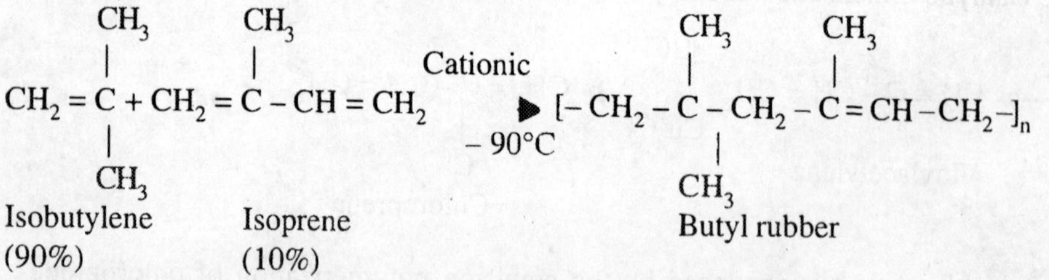

Butyl rubber has good chemical inertness, low gas permeability, less sensitive to oxidative aging, better ozone resistance and good solvent resistance. It is used mainly for the manufacture of inner tubes of tyres.

(d) Silicone rubber

Silicone rubber (polymethyl siloxane) is an example for inorganic polymers. It is a chain of alternating silicon and oxygen atoms and each silicon atom carries two methyl groups. It has the following structure:

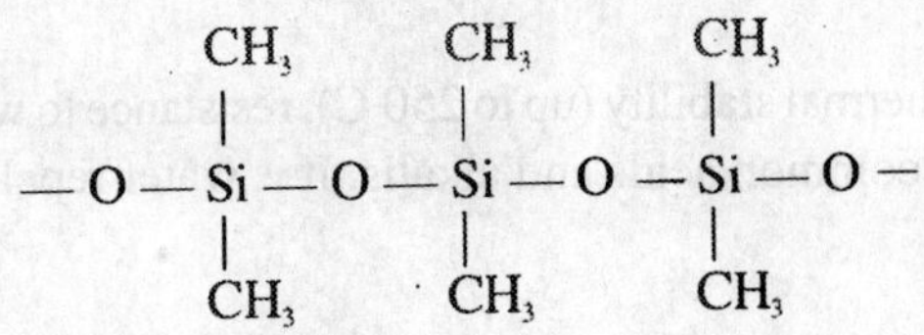

Preparation

Silicone rubber is formed by the condensation polymerization of dimethyl silanol, which is produced from dimethylchloro silane.

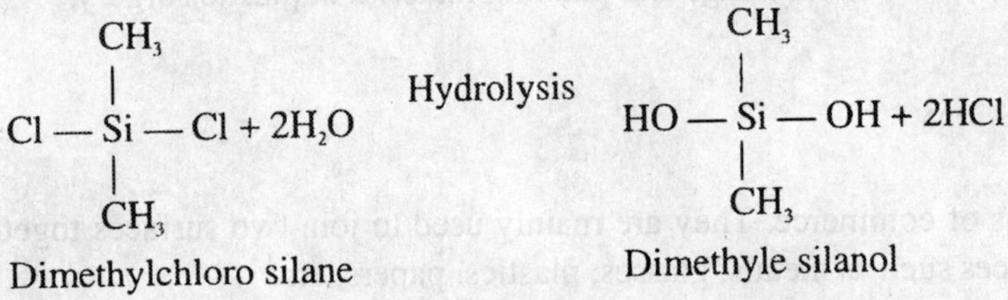

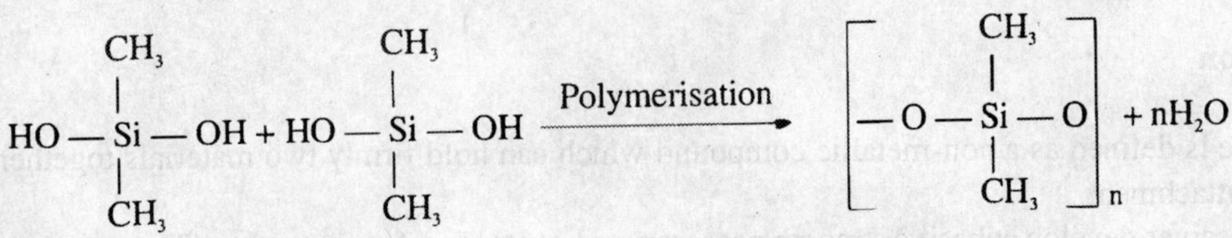

Vulcanization

Because of the absence of unsaturation in silicones, sulphur vulcanization is not possible. It is, therefore, vulcanized by using peroxides such as H_2O_2 or benzoyl peroxide. Silicon chains are cross linked through methylene bridges.

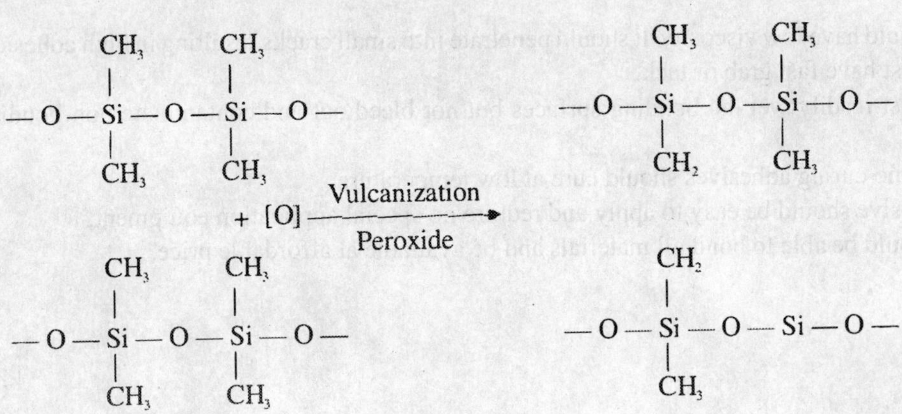

Properties

Silicone rubber possesses high thermal stability (up to 250°C), resistance to weathering and lubricating properties. It is not affected by common acids and alkalis. It is water repellent.

Applications

It is used for:

(i) Industrial and domestic gaskets
(ii) High temperature cable insulation and safety glass interlayers.
(iii) Water proof surfaces and clothing
(iv) Tyres of fighter air-crafts
(v) Artificial heart valves, transfusion tubings and padding material in plastic surgery.

7.13. ADHESIVES

Adhesives are important items of commerce. They are mainly used to join two surfaces together. They join a variety of substances such as metals, glasses, plastics, papers, etc.

Definition

Adhesive is defined as a non-metallic compound which can hold firmly two materials together by surface attachment.

Adhesives develop cohesive strength necessary to hold two surfaces together. They can transfer load from one substance to another. The bonded materials will have the properties of the substances.

Adhesion offers many advantages, over soldering, riveting, welding, screwing, etc. It can be carried out more easily, quickly and economically as well.

Requirements of Adhesives

- It should have low viscosity. It should penetrate into small cracks resulting in high cohesion.
- It must have fast grab or tack.
- It must readily wet the bonding surfaces but not bleed out and contaminate non-bonding areas.
- Thermo curing adhesives should cure at low temperature.
- Adhesive should be easy to apply and require no special application equipment.
- It should be able to bond all materials and be available at affordable price.

Surface cleaning

To ensure effective cohesion of adhesives, the surface of materials to be joined must be properly cleaned before the adhesives are applied. With paper and wood, the surfaces are ordinarily used in the "as received" condition. Layers of grease, oil and lubricants are removed by means of suitable solvents. Loose dirt and scales on metal surfaces can be removed by brushing. In many cases, the surfaces are rubbed by emery papers or sand blasted to remove deposits on the surfaces. In addition to surface cleaning, the primers are often applied before the application of actual adhesives in some cases to influence the resistance to aging and corrosion.

There are mainly two types of adhesives: (a) Natural and (b) Synthetic.

Natural Adhesives

Gum, glue, casein, starch, etc., are the important natural adhesives. These are not all purpose adhesives, besides, they have lower strength.

Synthetic Adhesives

They are low molecular weight and soluble polymers called resins. The important adhesives of this class are PF, UF, MF, epoxy resins and silicones. These are soluble in suitable solvents and become thermosetting on curing. These resins have high strength, resistance to water, and corrosion. Further, they are unaffected by weather.

Epoxy resin (Araldite)

The trade names of epoxy resin are '**Araldite**' and '**Epon**'.

These were first synthesized in 1940s by step polycondensation mechanism. Epoxy resins are obtained by the condensation of excess **Epichlorohydrin** with **Bisphenol A,** catalysed by alkali as per the scheme, shown in Fig. 7.7.

These prepolymers are linear, low molecular weight and soluble liquids. These prepolymers are cross linked, (cured) at room temperature on mixing with diamines, dicarboxylic acids or acid anhydrides.

$$\text{Epoxy resin + Acid anhydride} \xrightarrow{\text{room temp}} \text{Cross linked (cured) epoxy resin}$$

Epoxy resins are highly resistant to water, solvents, acids and alkalies.

- They are widely used as structural adhesives because of their excellent chemical resistance and good adhesion.
- They are used for laminating materials.
- They are also used to impart crease and shrinkage resistance to fabrics.

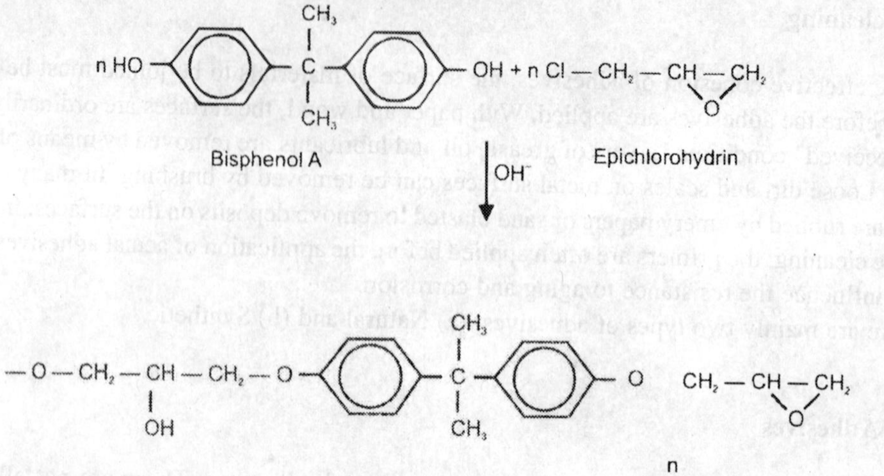

Fig. 7.7. Epoxy prepolymers

7.14. POLYMER COMPOSITES

Composite materials may be defined as materials composed of at least two distinctly dissimilar materials acting in harmony. The properties of the composite system are not attainable by the individual components acting alone. A plastic composite, consists of a combination of a polymeric resin matrix with reinforcing fibrous material. The combination yields a unique class of materials which provide a user with tailorability, that allows properties to be specifically tailored to the requirement.

When compared to traditional materials, composites are found to have higher strength per unit weight, which can be translated directly into weight savings. These reductions in weight, in turn, result in more efficient structures, reduced energy costs, and reduced material cost. Composites also offer many advantages such as resistance to fatigue and corrosion, low cost of fabrication, tailored thermal expansion characteristics and thermal conductivity, and design flexibility by the proper choice of reinforcement locations to design requirements. Biopolymeric composites, such as feathers and wood have been available for thousands of years and fiberglass reinforced polyester was the earliest man made polymer composite.

Polymer composites are produced by suitably bonding a fiber material with a polymer resin matrix and curing the same under pressure and heat. The reinforcements could be in different forms such as short chopped fibers, continuous filaments, twisted strands, mats, woven fabrics, etc.

The common resin matrix used in polymer composites includes polyesters, epoxyresins, phenolic resins, silicone resins, melamine resins, vinyl derivatives and polyimide. Phenolic resins are used wherever resistance to high temperature is demanded. Epoxy resins impart high mechanical properties and silicone resins impart excellent electrical and thermal properties.

The various reinforcement fibers used in polymer composites are glass fiber, boron filaments, carbon / graphite fiber, aramid (aromatic polyamide) fiber, alumina fiber, polycrystalline metal oxide fibers, silicon carbide fiber, etc.

The Hand Lay-up Technique of producing polymer composites, consists of the following steps. The mould is given a thin coating of silicone oil or wax to prevent the final article from sticking to the mould. The mould is then coated with a resin matrix and over which a precut glass fiber cloth or mat is laid. Another layer of the resin coating is given over the glass cloth. Rollers are used to press the glass cloth on the resin uniformly and also to remove the entrapped air bubbles. Alternate layers of resin and glass cloth are laid in a similar sequence until the required thickness is built up. The whole set up is then cured at either normal temperature or elevated temperature. The so formed composite is removed from the mould and subjected to trimming and finishing.

Kevlar [–NH–C_6H_4–NH–CO–C_6H_4–CO–]$_n$

Kevlar is an aramid, a term invented as an abbreviation for aromatic polyamide. The chemical composition of Kevlar is poly para-phenyleneterephthalamide, and it is more properly known as a para-aramid. Developed at DuPont by Stephanie Kwolek's in 1965, this high strength material was first commercially used in the early 1970s as a replacement for steel in racing tires. Typically it is spun into ropes or fabric sheets that can be used as such or as an ingredient in composite material components. It was the first organic fiber with sufficient tensile strength and structural rigidity to be used in advanced composites.

Kevlar is synthesized in solution of *N*-methyl-pyrrolidone and calcium chloride from the monomers 1,4-phenylene-diamine (para-phenylenediamine) and terephthaloyl chloride through a condensation reaction with the liberation of HCl as a byproduct.

Properties

Kevlar is a very strong material - about five times as strong as steel. Kevlar fibers possess the following properties:

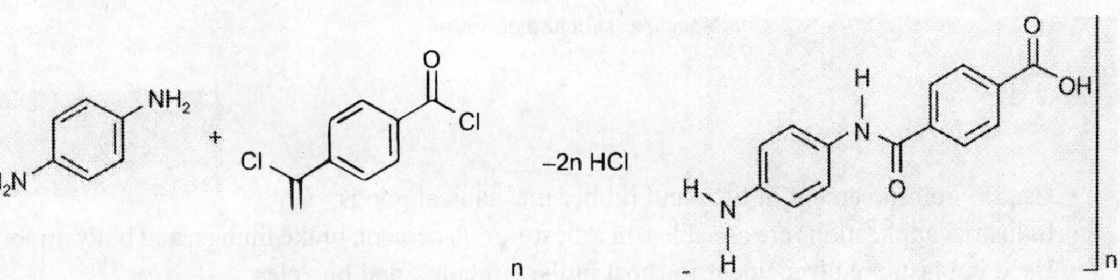

- High tensile strength (five times stronger per unit weight than steel);
- High modulus of elasticity;
- Very low elongation up to breaking point;
- Low weight;
- High chemical inertness;
- Very low coefficient of thermal expansion;
- High fracture toughness (impact resistance);
- High cut resistance;
- Textile processibility;
- Flame resistance.

The polymer owes its high strength to the many inter-chain hydrogen bonds. These intermolecular hydrogen bonds are formed between the carbonyl groups and N*H* centers. Additional strength is derived from aromatic stacking interactions between adjacent strands. These interactions have a greater influence on Kevlar than the van der Waals interactions and chain length that typically influence the properties of other synthetic polymers and fibers. Kevlar's structure consists of relatively rigid molecules which tend to form mostly planar sheet-like structures.

Molecular structure of Kevlar: bold represents a monomer unit, dashed lines indicate hydrogen bonds.

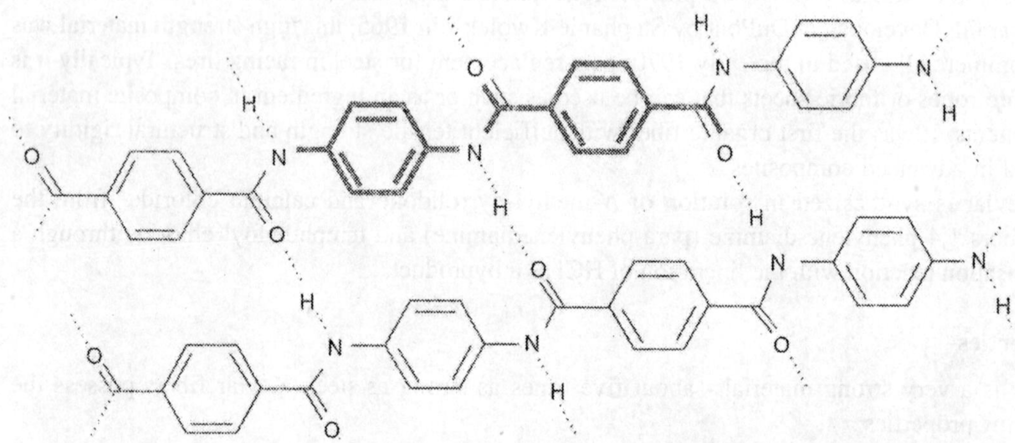

Molecular structure of Kevlar

Uses:

- Used as reinforcement in tires and rubber mechanical goods.
- Industrial applications are as cables, in asbestos replacement, brake linings, and body armor.
- Used in plastic reinforcement for boat hulls, airplanes, and bicycles.
- Kevlar is well-known component of personal armor such as combat helmets, ballistic face masks, etc.
- Used as an inner lining for some bicycle tires to prevent punctures.
- The fiber is used in woven rope and in cable.
- Aramid fibers are widely used for reinforcing composite materials, often in combination with carbon fiber and glass fiber. The matrix for high performance composites is usually epoxy resin. Typical applications include monocoque bodies for F1 racing cars, helicopter rotor blades, tennis, table tennis, badminton and squash rackets, kayaks, cricket bats, and hockey sticks.

Disadvantages

Like all good things, Kevlar also has a few disadvantages.

- The fibers themselves absorb moisture, so Kevlar composites are more sensitive to the environment than glass or graphite composites.
- Although tensile strength and modulus are high, compressive properties are relatively poor.
- Kevlar is also very difficult to cut. Needs special scissors for cutting dry fabric and special drill bits for drilling cured laminates.

Carbon Fiber (Graphite Fiber, CF)

Carbon fiber is a material consisting of extremely thin fibers (0.005–0.010 mm in diameter) and composed mostly of carbon atoms. The carbon atoms are bonded together in microscopic crystals that are more or less aligned parallel to the long axis of the fiber. Several thousand carbon fibers are twisted together to form a yarn which may be used by itself or woven into a fabric. Carbon fiber can be combined with a plastic resin and moulded to form carbon fiber reinforced plastics to provide a high strength-to-weight ratio material. The density of carbon fiber is also considerably lower than the density of steel, making it ideal for applications requiring low weight.

Synthesis

Each carbon filament is produced from a precursor polymer. The precursor polymer is commonly rayon, polyacrylonitrile (PAN) or petroleum pitch. For synthetic polymers such as rayon or PAN, the precursor is first spun into filaments, using chemical and mechanical processes. After drawing or spinning, the polymer fibers are then heated to eliminate non-carbon atoms (carbonization), producing the final carbon fiber. The carbon fibers may be further treated to improve handling qualities, and then wound on to spindles.

Carbon fibers are prepared from PAN by heating the spun PAN filaments to approximately 300°C in air, which breaks many of the hydrogen bonds and oxidizes the material. The oxidized PAN is then heated to approximately 2000°C, in a furnace, in an inert atmosphere of argon gas, which induces carbonization of the material. The resultant fibers usually contain 93–95% of carbon.

Structure

The atomic arrangement in carbon fiber is similar to that of graphite, consisting of sheets of carbon atoms arranged in a regular hexagonal pattern. The difference lies in the way these sheets interlock. In graphite sheets are stacked parallel to one another in regular fashion. The intermolecular forces between the sheets are relatively weak van der Waals forces, giving graphite its soft and brittle characteristics. Depending upon the precursor to make the fiber, carbon fiber may be turbostratic or graphitic, or have a hybrid structure with both graphitic and turbostratic parts present. In turbostratic carbon fiber the sheets of carbon atoms are haphazardly folded, or crumpled, together.

Carbon fibers derived from PAN are turbostratic, whereas carbon fibers derived from pitch are graphitic after heat treatment at temperatures exceeding 2200°C.

Properties and uses

- The properties of carbon fiber such as high tensile strength, low weight, and low thermal expansion make it very popular in aerospace, civil engineering, military, and motorsports, along with other competition sports.
- Carbon fiber is very strong when stretched or bent, but weak when compressed or exposed to high shock (eg. a carbon fiber bar is extremely difficult to bend, but will crack easily if hit with a hammer).
- Carbon fiber is used as reinforcing element in carbon fiber reinforced polymers. Non-polymer materials can also be used as the matrix for carbon fibers.
- Moulding a thin layer of carbon fibers significantly improves fire resistance of polymers or thermoset composites because a dense, compact layer of carbon fibers efficiently reflects heat.
- It has also been used in the field of medical treatment to treat severe burns as carbon fiber threads in skin grafts.
- Carbon fiber filament yarns are used in several processing techniques: the direct uses are for filament winding, pultrusion, weaving, braiding, etc.

However, carbon fibers are relatively expensive when compared to similar materials such as fiberglass or plastic.

Carbon fiber-reinforced polymer or carbon fiber-reinforced plastic (CFRP or CRP)

Carbon fiber-reinforced polymer or carbon fiber-reinforced plastic is a very strong, light, and expensive composite material or fiber reinforced polymer. The composite material is commonly referred to by the name of its reinforcing carbon fiber. The polymer mostly used is epoxy, but other polymers, such as polyester, vinyl ester or nylon are sometimes used.

Carbon Fiber Reinforced Polymers (CFRP) materials usually have laminate structure, providing reinforcing in two perpendicular directions.

Properties

- Light weight
- High strength-to-weight ratio
- Very high modulus elasticity-to-weight ratio
- High fatigue strength
- Good corrosion resistance
- Very low coefficient of thermal expansion

- Low impact resistance
- High electric conductivity

Applications

- In aerospace and automotive fields, as well as in sailboats.
- In modern bicycles and motorcycles
- In consumer goods such as laptops, tripods, fishing rods, archery equipment, racket frames, stringed instruments bodies, guitat strings, golf clubs. and pool/billiard cues.

7.15. CONDUCTING POLYMERS

Organic polymers are normally used in electrical and electronic applications as insulators, where advantage is taken of their excellent insulating properties. However, the curiosity of chemists in producing polymers having unusual electric and electronic properties, has lead to the synthesis of new class of polymers, known as conducting polymers. Heeger, Macdiamid and Shirakawa, in 1977, demonstrated for the first time that electrical conductivity of polyacetylene can be increased 13 folds of magnitude by doping with electron acceptors and donors. Norman and others subsequently achieved conductivity as high as that of copper metal in polyacetylene. Now it has been established that doping of polymers having pi-back bone, with less than 10% chemical species results in dramatic changes in their electrical, electronic, magnetic and optical properties. These conducting polymers constitute a new family of materials with highly delocalized pi-electron system.

Conducting polymers are found to be highly promising materials for various applications with inherent advantages such as flexibility, ease of fabrications, stability, ease of processability, and above all with the low cost conducting polymers are considered to be better than conventional conducting inorganic materials.

Definition

"An organic polymer with highly delocalized pi-electron system, having electrical conductance of the order of a conductor is called a conducting polymer".

The concept of conductivity of conjugated polymers is now extended from polyacetylene to a number of conjugated polymers such as polypyrrole, polythiophene, polyphenylene, polyphenylene sulphide and polyaniline. Chemical structures of these compds are shown in Fig. 7.8.

Synthesis

An organic polymer can be converted into a conducting polymer if it has

- linear structure
- extensive conjugation in polymeric back bone (pi-back bone)

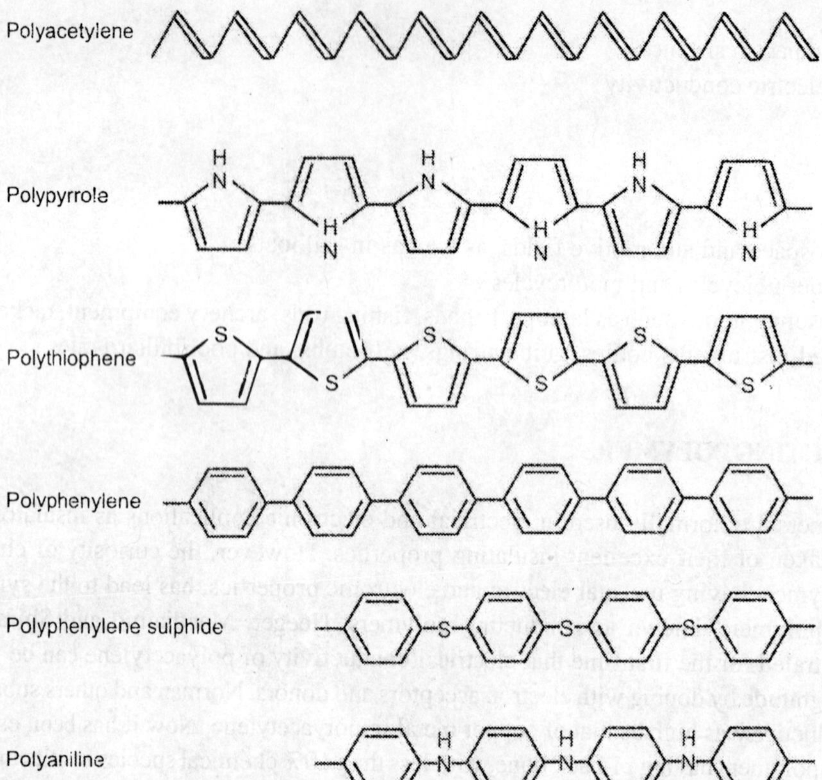

Fig. 7.8. Chemical structures of some important conducting polymers.

The conducting polymers are synthesized by doping, in which charged species are introduced in organic polymers having pi-back bone. The important doping reactions are:

1. Oxidative doping (p-doping)
2. Reductive doping (n-doping)
3. Protonic acid doping (p-doping)

1. Oxidative doping (p-doping, Fig. 7.9): In this process, pi-backbone of a polymer is partially oxidized using a suitable oxidizing agent. This creates positively charged sites on polymer back bone, which are current carriers for conduction.

The oxidizing agents used in p-doping are iodine vapor, iodine in CCl_4, HBF_4, perchloric acid and benzoquinone.

Mechanism of conduction: The removal of an electron form the polymer pi-back bone using a suitable oxidizing agent leads to the formation of delocalized radical ion called **polaron.**

A second oxidation of a chain containing polaron, followed by radical recombination yields two charge carriers on each chain. The positive charge sites on the polymer chains are compensated by anions I_3^- formed by the oxidizing agent during doping.

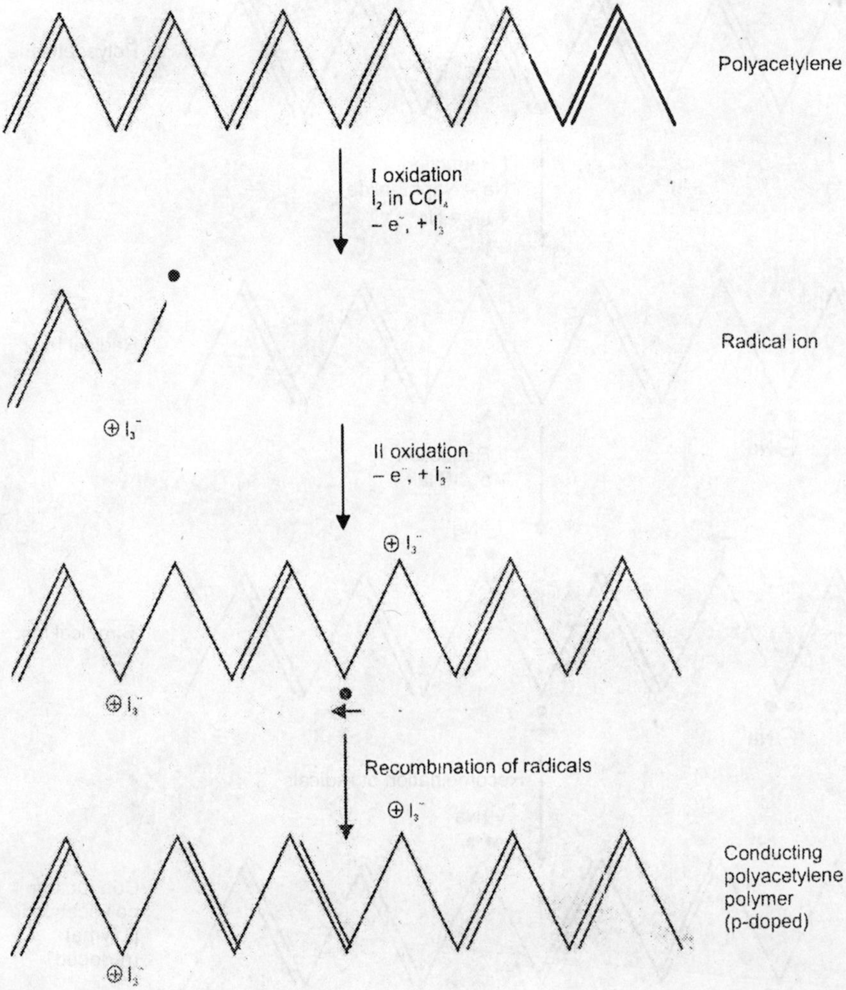

Fig. 7.9. Reactions of p-doping of polyacetylene

The delocalized positive charges on the polymer chain are mobile, not the dopant anions. Thus, these delocalized positive charges are current carriers for conduction. These charges must move form chain to chain as well as along the chain for bulk conduction. On doping polyacetylene using iodine in CCl_4, for partial oxidation, the conductivity increases from 10^{-5} S.cm^{-1} to $10^3 - 10^5$ S.cm^{-1}.

2. Reductive doping (n-doping, Fig. 7.10): In n-doping, pi-back bone of a polymer is partially reduced by a suitable reducing agent. This creates negatively charged sites on polymer backbone. These negatively charged sites are responsible for the conduction. The most commonly used reducing agent for partial reduction of polymer back bone is a solution of sodium naphthalide in tetra hydro furan.

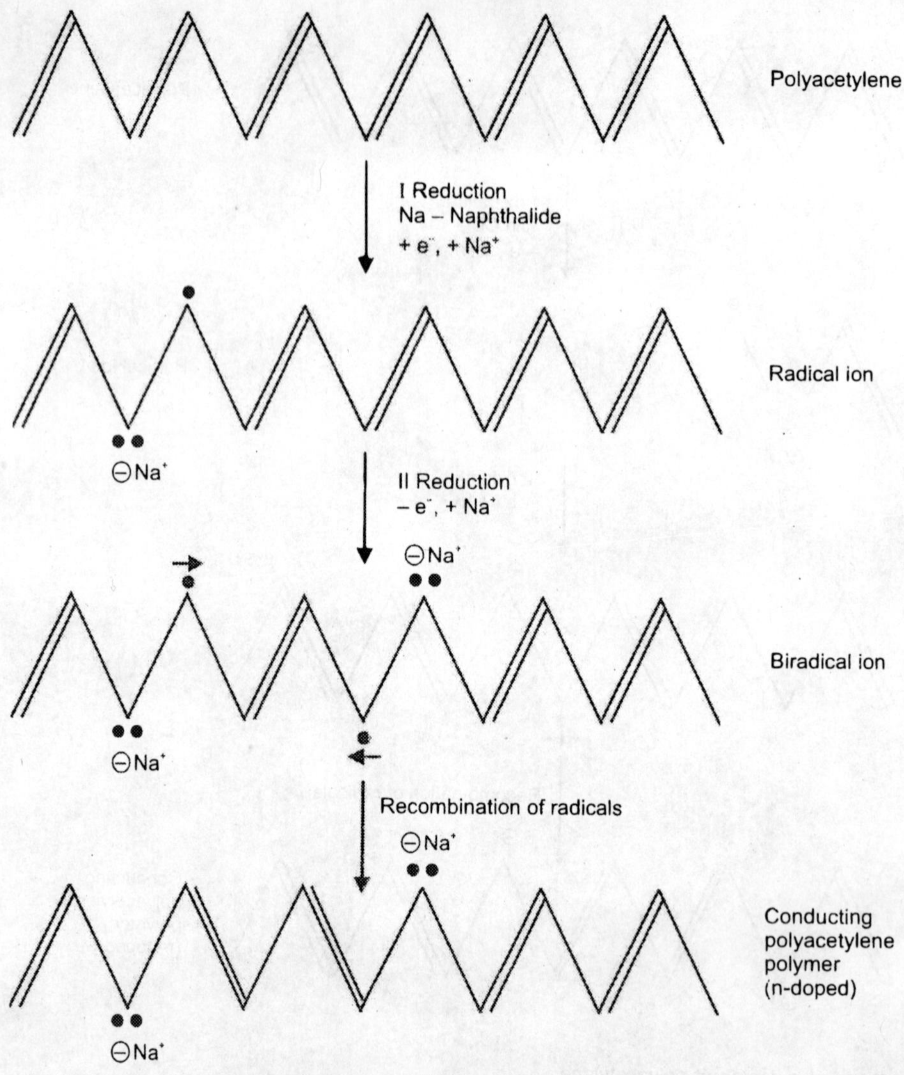

Fig. 7.10. Reactions of n-doping of polyacetylene

Mechanism of conduction: The addition of an electron to the polymer back bone by using a reducing agent generates a radical ion or polaron. A second reduction of chain containing polaron, followed by the recombination of radicals yields two charged (–ve) carriers on each chain. These charge sites on the polymer chains are compensated by cations (Na^+ ions) formed by the reducing agent.

3. Protonic acid doping (Fig. 7.11): The synthesis of conducting polyaniline is a typical example of this type of doping technique. In this technique, current carrying charged species (–ve / +ve) are created by the protonation of imine nitrogen.

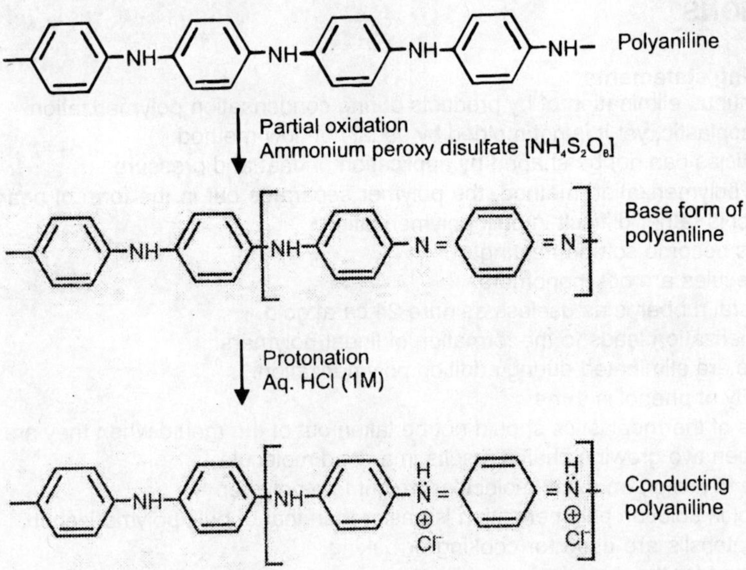

Fig. 7.11. Reactions of protonic acid doping of polyaniline

Polyaniline is partially oxidized first, using a suitable oxidizing agent, into a base form of aniline which contains alternating reduced and oxidized forms of aniline polymer back bone. This base form of aniline when treated with aqueous HCl (1M), undergoes protonation of imine nitrogen atom, creating current carrying charged sites (+ve) in the polymer back bone. These charges are compensated by the anions (Cl^-) of the doping agent, giving the corresponding salt.

This type of protonic acid doping of polyaniline results in an increase of conductivity by approximately 9–10 orders of magnitude.

Applications

Conducting polymers are highly promising materials to be used in electric and electronic applications. Some of the applications are:

- as electrode material for commercial rechargeable batteries, for higher power to weight ratio (coin type batteries)
- as conductive tracks on printed circuit boards.
- as sensors-humidity sensor, gas sensor, radiation sensor, biosensor for glucose, galactose, etc.
- in electrochromic display windows.
- in information storage devices.
- as film membranes for gas separations.
- in light emitting diodes.
- In fuel cells as the electro catalytic materials.

REVIEW QUESTIONS

I. Justify the following statements:

1. There is continuous elimination of by products during condensation polymerization.
2. Teflon is thermoplastic, yet it is not molded by usual hot-flow method.
3. Thermoset articles can not be shaped by application of heat and pressure.
4. In suspension polymerization method, the polymer separates out in the form of pearls or beads.
5. Thermal control is rather difficult in bulk polymerization.
6. Thermoplastics become soft on heating.
7. All simple molecules are not monomers.
8. The virgin natural rubber is as useless as pure 24 carat gold.
9. Addition polymerization leads to the formation of linear polymers.
10. No by products are eliminated during addition polymerization.
11. The functionality of phenol is three.
12. Molded articles of thermoplastics should not be taken out of the mould when they are still hot.
13. Collision between two growing chains results in a dead molecule.
14. It is impossible to assign an exact molecular weight to a polymer.
15. Thermal control in solution polymerization is easier than that of bulk polymerization.
16. Teflon coated utensils are used for cooking or baking.
17. Plexiglass is used for the manufacture of lenses.
18. Fillers are used in compounding of plastics
19. Elastomers undergo reversible elongation.
20. Buna – S is a copolymer.
21. Free radical polymerization is carried out in the absence of oxygen.
22. Polymers formed by anionic polymerization are called living polymers.

II. Answer the following:

1. Define the following terms and give examples:
   Monomer, polymer, polymerization reaction, functionality, degree of polymerization.
2. Classify the polymers based on:
   (i) Thermal behavior
   (ii) Mechanism of polymerization
   (iii) Properties
3. Explain the free radical polymerization mechanism of a vinyl compound.
4. Define and differentiate addition and condensation polymerization. Give suitable examples.
5. Give a brief account of glass transition temperature.
6. Explain how the structure of a polymer influences the following properties:
   (i) Tensile strength
   (ii) Crystallinity
   (iii) Plasticity
   (iv) Chemical resistivity.
7. What is the difference between resins and plastics? Give examples.
8. Define and differentiate thermoplastics and thermosettings.
9. Explain how the following polymers are produced and mention their applications.
   (i) Teiflon,
   (ii) Plexiglass
   (iii) Polyurethane,
   (iv) HDPE,

(v) P-F resin
(vi) Polycarbonates

10. Give an account, on Kevlar and carbon fibres.
11. What are elastomers? Write the structure for natural rubber and gutta percha. What are the differences between them?
12. Explain how the following elastomers are obtained. Mention their applications
    (i) Buna – S
    (ii) Neoprene
    (iii) Butyl rubber
    (iv) Silicon rubber
13. Write a note on adhesives.
14. Describe the following methods of polymerization. Point out their merits and demerits.
    (i) Bulk
    (ii) Solution
    (iii) Suspension
    (iv) Emulsion
15. Write notes on the following:
    (i) Polymer composites
    (ii) Conducting polymers.
16. Derive expressions for number average molecular.
17. Explain the cationic mechanism of addition polymerisation.
18. Explain anion mechanism of addition polymerisation.

CHAPTER 8

Water Treatment

8.1. INTRODUCTION

Water is the most wonderful and precious gift of nature to all the living beings on the earth. It is the most abundant and useful compound on the earth. It covers nearly 72% of the earth's surface. Water gets evaporated continuously from ocean, rivers and lakes, goes into the atmosphere as clouds and comes down to the earth's surface in the form of rain. During the course of its downward journey, the rain water absorbs many impurities such as gases, dust and other impurities present in the atmosphere. After reaching the earth's surface, it flows first as stream and then as rivers. The flowing river water picks up many mineral salts, silt and clay from land and also effluents discharged by the industries, domestic, social and public establishments. At the end of its long journey, finally it merges into the sea from where it has originated.

Almost all water on earth, 97.2%, is locked in the ocean. About 2.1% is in the form of ice and glaciers. The fresh water, therefore available in rivers, lakes and under ground, amounts only to 0.6%. The remaining 0.1% is in the brine wells and salty water.

Water is the most widely used chemical compound on earth. Among the three essentials of life, i.e., air, water and food, water occupies the second position in the order of priority. It is not only an essential life line for human beings, but also for plants and animals. Human body contains about 65.0% by weight of water in the form of blood and other body fluids. The average intake of water per day by an individual is about 2.5 liters.

Water is needed for many domestic uses such as drinking, cooking and washing. In addition, a large quantity of water is consumed in many industrial processes, agricultural and power production. Health and wealth of a nation largely depends on the availability and utilization of quality water.

8.2. SOURCES OF WATER

The sources of water are classified as **surface water** and **underground water.**

Surface water

1. **Rain water** is supposed to be the purest form of water. However, it dissolves considerable amount of gases and suspended solid particles from the atmosphere during its downward journey through it.
2. **River water** contains dissolved minerals of the soils such as chlorides, sulfates, bicarbonates of sodium, calcium, magnesium and iron. It also contains the organic matter, derived from the decomposition of plants and small particles of sand and soil in suspension. Thus, river water is with considerable amounts of dissolved as well as suspended impurities.
3. **Lake water** has more or less a constant chemical composition. It contains less of dissolved minerals, but quite a high quantity of organic matter.
4. **Sea water** is the most impure form of natural water. It contains on an average of about 3.5% of dissolved salts, out of which about 2.6% is sodium chloride. Other salts present are sulfates of sodium; bicarbonates of potassium, magnesium and calcium; bromides of potassium and magnesium and a number of other compounds.

Underground water

Spring and well water are from the underground water sources, and are, in general, clearer in appearance due to the filtering action of the soil, but contain more of the dissolved salts. Usually, underground water is of high organic purity.

8.3. IMPURITIES IN WATER

The water found in nature is never pure and contains a large number of impurities in varying amounts. The impurities associated with water largely depend upon its source. The major types of impurities found in water are of the following types:

(i) Dissolved gases

The water contains mainly carbon dioxide and oxygen as dissolved gases. Some water may contain ammoni and sulfur compounds such as hydrogen sulfide also as dissolved gases, which impart foul smell to water.

(ii) Dissolved solids

The soluble salt impurities present in water include salts of calcium, magnesium, sodium in various soluble salt forms. Oxides of manganese, iron, lead and arsenic may also present in water.

(iii) Suspended impurities

Suspended impurities are the dispersion of solid particles, which can be removed by filtration or settling. They are of two types – inorganic and organic. Clay, silica, oxides of iron and manganese are of inorganic type of suspended impurities. Wood pieces, disintegrated particles of dead animals, leaf, fishes, bacteria, algae, protozoa, etc., are of organic origin. The suspended impurities impart turbidity and color to water.

(iv) Microscopic matter

Many pathogenic bacteria and microorganisms are also present in water. They are the main causes for the water borne diseases.

Hardness of water: Hardness of water is due to the presence of soluble salts of calcium, magnesium and other heavier metals in water. Hard water does not give lather easily with soap, but produces a white scum or precipitate with soap. This is due to the formation of insoluble soaps of calcium and magnesium when the calcium and magnesium ions react with soap. (Soaps are sodium and potassium salts of higher fatty acids such as oleic acid, palmitic acid and stearic acid).

$$\underset{\text{Sodium stearate}}{2C_{17}H_{35}COONa} + Ca^{2+} \rightarrow \underset{\text{precipitate}}{(C_{17}H_{35}COO)_2Ca\downarrow} + 2Na^+$$

$$\underset{\text{Sodium stearate}}{2C_{17}H_{35}COONa} + Mg^{2+} \rightarrow \underset{\text{precipitate}}{(C_{17}H_{35}COO)_2Mg\downarrow} + 2Na^+$$

Therefore, soap can produce lather only after all the hardness causing ions are precipitated as insoluble soap. Hence, it requires large quantity of soap to produce lather. Thus, water which does not produce lather readily with soap is called hard water. On the other hand, water which produces lather easily with soap is called soft water.

Hardness of water can be classified as temporary hardness and permanent hardness.

Temporary hardness (Carbonate hardness): Temporary hardness is caused by the presence of dissolved bicarbonates of calcium and magnesium. This hardness can be removed by mere boiling of water, when bicarbonates are converted into insoluble carbonates.

$$Ca(HCO_3)_2 \rightarrow CaCO_3\downarrow + H_2O + CO_2$$

$$Mg(HCO_3)_2 \rightarrow MgCO_3\downarrow + H_2O + CO_2$$

Permanent hardness (non-carbonate hardness): Permanent hardness is caused by the dissolved salts of calcium and magnesium, other than bicarbonates. They are chlorides, sulfates, nitrates, etc. Unlike temporary hardness, permanent hardness can not be removed by boiling. Removal of this hardness requires certain special chemical treatment methods.

The sum of temporary and permanent hardness is referred to as **total hardness** of water. The hardness is expressed in terms of equivalent calcium carbonate. The equivalent of $CaCO_3$ is given by,

$$\text{Equivalent of } CaCO_3 = \frac{(\text{Mass of Hardness causing salt}) \times (\text{Chemical equivalent of } CaCO_3)}{\text{Chemical equivalent of hardness producing salt}}$$

Table 8.1 gives the multiplication factor for converting various salts into equivalents of calcium carbonate.

Table 8.1. Calculation of equivalents of calcium carbonate

Dissolved salt	*Molecular mass*	*Chemical equivalent*	*Multiplication factor for converting into equivalents of $CaCO_3$*
$Ca(HCO_3)_2$	162	81	100 / 162
$Mg(HCO_3)_2$	146	73	100 / 146
$CaCl_2$	111	55.5	100 / 111
$MgSO_4$	120	60	100 / 120
$MgCl_2$	95	47.5	100 / 95
$MgCO_3$	84	42	100 / 84
$Mg(NO_3)_2$	148	74	100 / 148
$Ca(NO_3)_2$	164	82	100 / 164

The units normally used to express hardness are **parts per million (ppm)** or **milligrams per litre (mg/litre)** of calcium carbonate equivalents.

Parts per million (ppm) is the parts of calcium carbonate equivalents per million (10^6) parts of water.

Example 8.1. Calculate the temporary hardness and permanent hardness of a sample of water, which contains 16.2 mg of calcium bicarbonate, 29.2 mg of magnesium bicarbonate, 33.3 mg of calcium chloride, 18 mg of magnesium sulfate and 55 mg of sodium chloride per litre of water.

Solution:

Calcium carbonate equivalent of 16.2 mg of $Ca(HCO_3)_2$ $= (100/162) \times 16.2 = 10$ mg

Calcium carbonate equivalent of 29.2 mg of $Mg(HCO_3)_2$ $= (100/146) \times 29.2 = 20$ mg

Calcium carbonate equivalent of 33.3 mg of $CaCl_2$ $= (100/111) \times 33.3 = 30$ mg

Calcium carbonate equivalent of 18 mg of $MgSO_4$ = (100/120) × 18 = 15 mg.

Temporary hardness [due to $Ca(HCO_3)_2$ and $Mg(HCO_3)_2$] = 10 + 20

= 30 mg/litre = **30 ppm**

Permanent hardness (due to $CaCl_2$ and $MgSO_4$) = 30 + 15

= 45 mg/liter = **45 ppm**

Total hardness = 30 + 45 = **75 ppm**

(Sodium salts do not cause hardness).

8.4. BOILER FEED WATER AND BOILER PROBLEMS

Boiler feed water is the water supplied to the boiler for the generation of steam or hot water. Boilers are used in variety of manufacturing industries as well as in power generating units. In such industries water is obtained from rivers, lakes and wells. The water available from these sources contains varying amounts of different types of impurities. Use of impure water in the boilers leads to different types of boiler problems that impair the effective use of boilers and also affect the quality of steam produced. Boiler problems may also result in safety related issues. Some of the boiler problems that may be encountered when impure water is used in the boilers are discussed in the following sections.

(a) Scales and Sludge Formation

Of the various impurities found in boiler feed water, the most harmful are those which cause hardness. In the boiler, the feed water containing such salts is continuously evaporated and pure steam is drawn out leaving behind the salts in the boiler water. With the continuous evaporation of water from the boiler, the concentration of the salts gradually increases. When the concentration of the dissolved salts reach the super saturation level, the salts begin to separating out as insoluble precipitates. The precipitates so formed in the boilers during steam generation are known as scales and sludges, based on their physical properties.

Sludges

If the precipitates formed in the boiler are soft, loose and non-adherent, they are called sludges. They get collected at the cooler region of the boiler and do not adhere firmly on the walls of the boiler metal. Therefore they can be easily flushed out of the boiler by blow down operations. The sludges are formed by salts such as calcium phosphate ($Ca_3(PO_4)_2$), calcium chloride ($CaCl_2$), magnesium chloride ($MgCl_2$), magnesium carbonate ($MgCO_3$), magnesium sulfate ($MgSO_4$), etc.

Scales

If the precipitates formed in the boiler are hard, adherent and form a firm deposit on the inner walls of the boiler, then they are called scales. Once the scales are formed on the boiler metal, they cannot be removed easily. The so formed layer of metal salts on the boiler metal are non-conducting and act as an insulator layer between the boiler metal and the water and hinder the heat transfer from the boiler to water.

Scales are mostly formed by salts like carbonates ($CaCO_3$), sulfates ($CaSO_4$) and silicates ($CaSiO_3$, $MgSiO_3$).

(i) Carbonate Scales: The boiler feed water invariably contains residual carbonate hardness. During steam generation in low pressure boilers, the residual calcium bicarbonate undergoes decomposition into calcium carbonate that gets deposited as carbonate scale on boiler surface.

$$Ca(HCO_3)_2 \rightarrow CaCO\downarrow + H_2O + CO_2$$

(ii) Sulphate Scale: The solubilities of certain salts decrease with the increase in temperature. Calcium sulfate is one such salt having negative temperature coefficient of solubility. The saturation level of calcium sulfate in boiler ware reaches fast on heating and gets deposited as a hard scale on the hot boiler metal surface. Calcium sulfate scale is the most adherent and the most difficult to remove, once formed.

(iii) Silicate Scales: The silicates of calcium and magnesium have very low solubility in cold water and reach zero solubility in superheated water in high pressure boilers. Therefore, presence of even small quantities of silicates in boiler water results in the formation of calcium silicate and magnesium silicate scales. The silicates scales are also deposited very firmly on the boiler metal and are very difficult to remove.

Harmful Effects of Scales

The deposition of an unıform, insulating, hard and adherent layer of scales on the inner surface of the boiler causes a number of adverse effects on the performance of the boiler. Some of the ill effects of scales are discussed in the following sections.

(i) Wastage of fuel: As the scales are poor conductors of heat, the heat supplied to the boiler is not efficiently transferred to water in the boiler. The boiler need to be heated to a much higher temperature than in the absence of scale in order to maintain a steady supply of steam. This overheating of the boiler results in wastage of valuable fuel. The extent of wastage of fuel increases with the thickness of the scale layer.

(ii) Decrease in boiler safety: The overheating of the boiler to maintain a steady supply of steam, makes the boiler metal soft, weak and often leads to distortion of the metal. The overheated boiler is, therefore, unsafe to bear the steam pressure, particularly, in high pressure boilers.

(iii) Boiler explosion: The coefficient of thermal expansion of the scale is very low compared to that of boiler metal. Overheating of boiler, therefore, causes uneven expansion of scale and metal surface, resulting in the formation of cracks in the scale layer. Water enters these cracks, comes in contact with the overheated metal surface and gets converted into steam at once. Formation of large quantity of steam at once, builds up high pressure inside the boiler which may result in boiler explosion.

(iv) Decrease in efficiency: The deposition of scales in boiler tubes, valves, etc., may clog them partially, leading to decrease in boiler efficiency.

Prevention of boiler scales

The scale formation is mainly due to the presence of salts of calcium and magnesium ions. Therefore, the scale formation in the boilers can be prevented if these metal ions are eliminated from the boiler feed water. The scale forming impurities can be eliminated by subjecting the boiler feed water to suitable treatments. These treatments will be discussed little latter in this chapter.

(b) Priming and Foaming

Priming

Priming is defined as the process of wet steam formation in the boiler. This process takes place when impure water boils violently in a boiler and some tiny droplets of water containing the dissolved impurities are carried along with the steam. The process of carrying the impurities along with the steam is known as 'carry over'.

Causes: Priming is mainly caused due to:

- Presence of suspended as well as dissolved impurities in water.
- Formation of scale and sludge in the boiler.
- Non uniform and sudden heating of water.
- High water level in the boiler, i.e., excess water feeding.
- Defective boiler design.
- Excessive foaming.

Harmful effects:

- The dissolved salts carried by wet steam get deposited on turbine blades and other machine parts. These deposits may cause corrosion of turbine blades and other machine parts; thereby decreasing their life as well as efficiency.
- The deposition of carried over impurity salts in valves, tubes, etc., results in decreased steam efficiency and corrosion of those parts.

Preventive measures: The priming can be minimized with the following measures:

- Using the properly designed boiler that can maintain adequate and proper boiling.
- Maintaining an optimum water level in the boiler.
- Feeding the water free from impurities.
- Periodical removal of concentrated salts and deposited sludges from the boiler by blow down operations.
- Regulating the rate of steam generation and demand for steam.

Foaming

Foaming is defined as the production of stable and persistent foam or bubbles on the surface of boiling water in the boiler. The stable bubbles containing the dissolved salts are carried along with the steam and result in priming. Thus both the priming and foaming occur together in the boiler.

Causes:

- The presence of oil and grease impurities in water.
- The residual alkalinity in the treated water, for example, sodium bicarbonate ($NaHCO_3$) from lime soda treatment. The oil is converted into soap under the prevailing alkalinity of water. They reduce the surface tension of water and promote excessive foaming.

Harmful effects: As foaming leads to priming, the harmful effects of priming are observed due to foaming also.

Preventive measures:

- Addition of anti-foaming agents such as castor oil, polyamides, etc.
- Addition of sodium aluminate which removes oil and grease.
- Feeding water free from foam causing impurities.

(c) Boiler Corrosion

The corrosion is a process of destruction of metal under the influence of the environment by chemical or electrochemical attack. If untreated water is fed into the boiler, it often leads to corrosion of boiler metal and tubes. Boiler corrosion is a serious problem encountered in steam boilers. The following are the main causes of boiler corrosion.

Presence of dissolved oxygen

Water invariably contains dissolved oxygen that can be one of the main causes of boiler corrosion. The high temperature and pressure prevailing in the boiler, facilitates the attack of oxygen on boiler metal, i.e., iron or steel.

$$2Fe + 2H_2O + O_2 \rightarrow 2Fe(OH)_2$$

$$2Fe(OH)_2 + 1/2\ O_2 \rightarrow Fe_2O_3 .\ 2H_2O$$

The boiler corrosion caused by the presence of dissolved oxygen can be prevented by eliminating oxygen from the boiler water. Dissolved oxygen can be removed by the following methods:

(i) Chemical treatment: The addition of certain reducing agents (oxygen scavengers) to boiler water effectively removes dissolved oxygen.

$$N_2H_4 + O_2 \rightarrow 2H_2O + N_2$$

$$Na_2S + 2O_2 \rightarrow Na_2SO_4$$

$$2Na_2SO_3 + O_2 \rightarrow 2Na_2SO_4$$

Among the various reducing agents used, hydrazine is the ideal one as it leaves behind no sallts.

(ii) Mechanical de-aeration: Dissolved oxygen can also be removed by mechanical method. The water is made to trickle through a series of perforated plates fitted in a large chamber, called de-aerator, so that a large surface area of water is exposed. The de-aerator is heated externally and connected to a vacuum pump to remove oxygen.

Presence of dissolved carbon dioxide

Water always contains variable amounts of dissolved carbon dioxide. Further, the decomposition of residual bicarbonates also contribute for an increase in carbon dioxide level in boiler water. Dissolved carbon dioxide produces carbonic acid (H_2CO_3), which attacks boiler metal and its parts, causing corrosion.

$$Mg(HCO_3)_2 \rightarrow MgCO_3 + H_2O + CO_2$$

$$CO_2 + H_2O \rightleftharpoons H_2CO_3$$

Carbon dioxide can be eliminated from boiler water by the addition of calculated amount of ammonia.

Presence of magnesium chloride

The presence of magnesium chloride produces hydrochloric acid due to hydrolysis.

$$MgCl_2 + 2H_2O \rightarrow Mg(OH)_2 + 2HCl$$

The hydrochloric acid attacks the boiler metal causing chemical corrosion.

$$Fe + 2HCl \rightarrow FeCl_2 + H_2$$

$$FeCl_2 + 2H_2O \rightarrow Fe(OH)_2 + 2HCl$$

Regenerated HCl leads to furhter corrosion of the boiler. Therefore, presence of small amount of magnesium chloride can cause continuous corrosion of the boiler.

Presence of residual alkali

Presence of residual alkali in the boiler water may lead to caustic embrittlement at the stressed regions of the boiler metal as discussed under corrosion chapter.

8.5 TREATMENT OF BOILER FEED WATER

In order to avoid the boiler problems it is desirable to treat the boiler feed water free of impurities. Boiler feed water is subjected to two types of treatments:

- External treatment and
- Internal treatment

a) External Treatment of Water

The treatment given to water before it enters the boiler is called the external treatment. The external treatment aims at removing / reducing the impurities responsible for boiler problem to a residual level that is harmless in the boiler water. One of the important methods of external treatment is **ion exchange process** or **demineralization process**.

Ion exchange (demineralization) process

An elegant way to soften water is to remove all the associated metal ions and their counter anions from water. This is accomplished with the efficient synthetic ion exchange resins.

The process of removal of all the associated ions from water by the application of ion exchange resins is known as demineralization of water. The utility of ion exchange resin rests with its ability to be used and reused, high efficiency, economical cost and long useful life.

Principle of ion exchange process: In ion exchange process the impure water is passed through cation exchange resin and anion exchange resin. Cation exchange resin is capable of exchanging cations present in water for H^+ ions, i.e., when the water passes through the cation exchanger, the cations present in the water as impurities are retained by the resin and H^+ ions are released into water. Anion exchanger resin is capable of exchanging anions present in water for OH^- ions, i.e., when water passes through the anion exchanger, the anions present in the water are retained by the resin and OH^- ions are released into water.

Function of cation exchanger: Commercially significant cation exchange resins are sulfonated copolymers of styrene and divinyl benzene (Fig. 8.1). These resins can be represented as RH and they exchange cations present in water.

$$2RH + Ca^{2+} \rightarrow R_2Ca + 2H^+$$

$$2RH + Mg^{2+} \rightarrow R_2Mg + 2H^+$$

$$RH + Na^+ \rightarrow RNa + H^+$$

$$RH + K^+ \rightarrow RK + H^+$$

The water coming out of the cation exchange resin is acidic as H^+ ions are introduced into water in place of metal ions.

Function of anion exchanger: The anion exchanger resins are coploymers of styrene and divinyl benzene containing active quaternary amino groups (Fig. 8.2). These resins are represented as ROH and they exchange anions present in water for OH^- ions.

$$ROH + Cl^- \quad RCl + OH^-$$

$$2ROH + SO_4^{2-} \rightarrow R_2SO_4^{2-} + 2OH^-$$

$$ROH + HCO_3^{2-} \rightarrow RHCO_3^- + OH^-$$

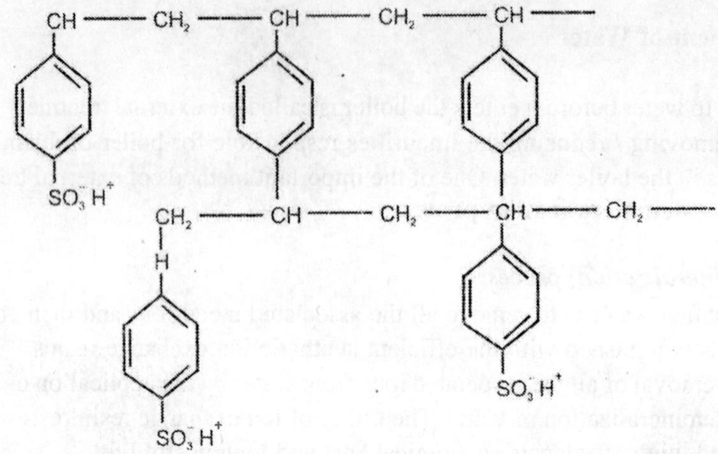

Fig. 8.1. Cation exchange resin.

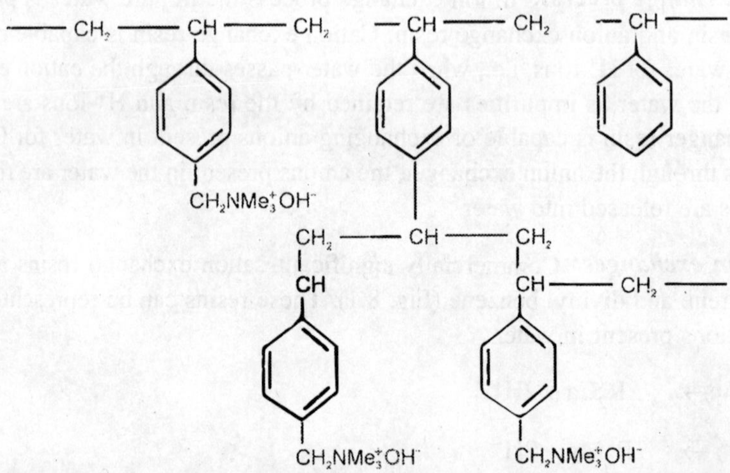

Fig. 8.2. Anion exchange resin

As cations are replaced by H^+ ions and anions by OH^-, the net effect is introduction of water in place of metal salts present in water.

$H^+ + OH^-$

The arrangement for demineralization of water is shown in Fig. 8.3.

Regeneration

When the resins get exhausted, they can be regenerated by passing HCl and NaOH solutions through cation and anion exchanger resins respectively.

In the cation exchanger:

$R_2Ca + 2HCl \qquad 2RH + CaCl_2$

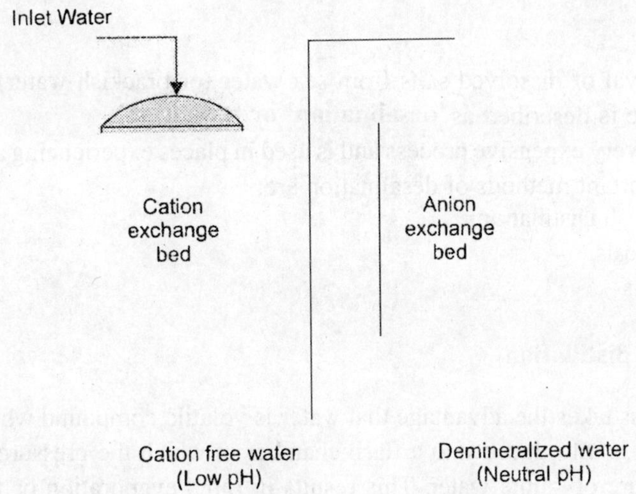

Fig. 8.3. Demineralization of water.

$$R_2Mg + 2HCl \qquad 2RH + MgCl_2$$

In the anion exchanger:

$$RCl + NaOH \qquad ROH + NaCl$$

$$R_2SO_4 + 2NaOH \qquad 2ROH + Na_2SO_4$$

Process

The raw water is first passed through the column containing cation exchanger resin, in which the cations present in water are retained and H^+ ions are released into water. The water coming out of the cation exchanger is acidic. The effluent from the cation exchanger column is then passed through the anion exchanger column. The column retains the anions present in water and releases OH^- ions.

8.6. DESALINATION

There are many countries in the world where the demand for fresh water far exceeds the availability of natural water. There are some countries like Kuwait, which have no fresh water, and have to look inevitably to the ocean for their fresh water demand. But high salt content of sea water prohibits its use for human consumption and other uses of water. There are three grades of water based on the level of dissolved salts, as given below.

(a) Fresh water with dissolved salts less than 1000 mg / l.
(b) Brackish water with dissolved salt between 1000 – 35,000 mg / l.
(c) Sea water with dissolved salts greater than 35,000 mg / l.

Definition

The process of removal of dissolved salts from sea water (or brackish water) to the extent, that water becomes usable is described as **'desalination' or 'desalting'**.

Desalination is a very expensive process and is used in places experiencing an acute shortage of fresh water. The important methods of desalination are:

(a) Multistage Flash Distillation
(b) Reverse osmosis
(c) Electrodialysis

(a) Multistage flash distillation

Principle: The process takes the advantage that water is volatile compound whereas salts are non-volatile. Saline water is introduced into a flash chamber in which the pressure is maintained less than the vapor pressure of saline water. This results in rapid evaporation or flash distillation of water. The steam on condensation produces fresh water. When this process is repeated several times, it is multi stage flash distillation. Latent heat of steam is used for the preheating of the saline water.

Process: A simplified multistage flash distillation unit is shown in Fig. 8.4.

The preheated brine is heated further by steam in a pre-heater. The hot brine (100°C) now passes into the first flash chamber. Since the pressure is lower in this flash chamber, a portion of the saline water flashes off (rapid evaporation) to form water vapor. The flashed water vapor condenses on the condenser coils, thus releasing its latent heat of condensation. The heat so released is utilized in heating the influent brine passing through the coils. The brine that remains after some of the water has been evaporated or flashed off is cooler. It then passes into the second flash chamber in which the pressure is little lower than in the first. A bit more of water evaporates and brine is cooled still further. In each successive stage, the brine becomes progressively more concentrated and is discharged from the last flash chamber. The fresh water from each chamber is collected and pumped off.

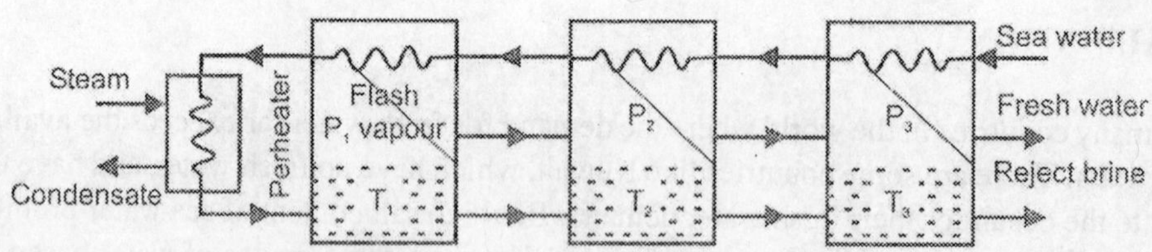

$T_1 > T_2 > T_3$
$P_1 > P_2 > P_3$

Fig. 8.4. Multistage flash distillation

The over all efficiency of multistage flash evaporator unit depends largely on the temperature of the influent saline water. Higher the temperature of influent saline water, greater is the efficiency of the unit. In practice, this temperature is however kept below 110°C to avoid some serious corrosion and scale problems.

(b) Reverse Osmosis

Principle: In reverse osmosis process, the water is separated from dissolved salts by means of a membrane which permits the passage of water through it but not the salts. If such a membrane is placed between brine and pure water, water has a natural tendency to flow through the membrane into the brine due to osmotic pressure. This natural process may be reversed by applying a pressure on the brine side higher than that of osmotic pressure, when fresh water tends to flow from brine into fresh water.

The process which reverses the natural spontaneous osmosis is called **"reverse osmosis"**

The principles of reverse osmosis are shown in Figure 8.5 in which a tube fitted with a semi permeable membrane and containing salt solution is partly immersed in water.

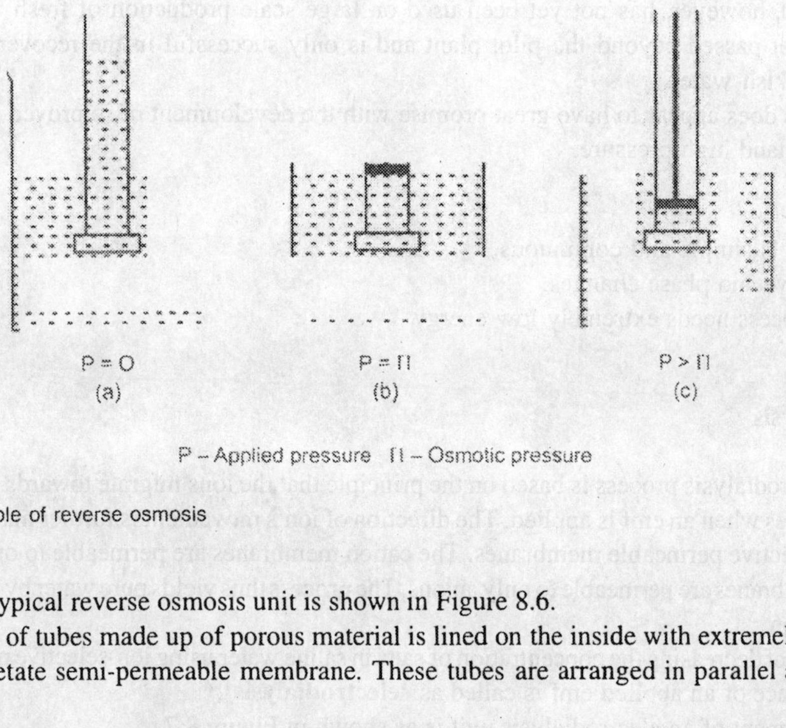

Fig. 8.5. Principle of reverse osmosis

Process: A typical reverse osmosis unit is shown in Figure 8.6.

A series of tubes made up of porous material is lined on the inside with extremely thin film of cellulose acetate semi-permeable membrane. These tubes are arranged in parallel array in fresh water.

Brackish water is pumped continuously at high pressure (> 25 atm) through these tubes. Water flows from brackish water into fresh water. The flow of water is proportional to applied pressure which in turn depends on the characteristics of the film. The film may rupture under excessive

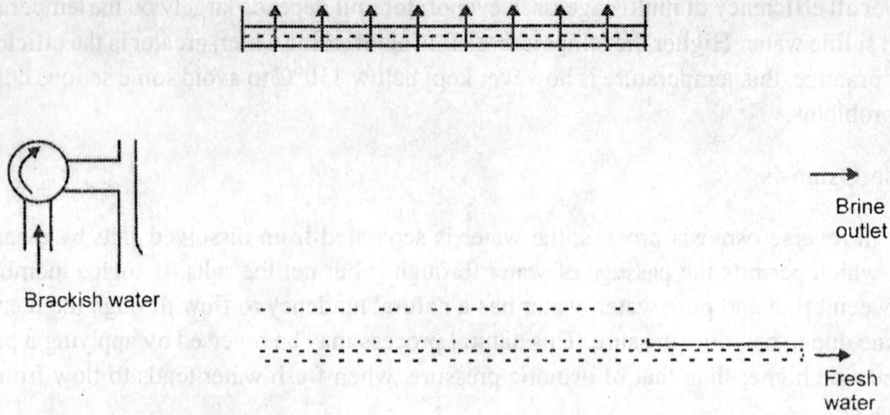

Fig. 8.6. Reverse osmosis method for desalination

pressure. Further, greater the number of tubes, larger is the surface area and, hence, more production of fresh water. Concentrated brine and fresh water are withdrawn through their respective outlets.

This method, however, has not yet been used on large scale production of fresh water. The technique has not passed beyond the pilot plant and is only successful in the recovery of fresh water from brackish water.

This process does appear to have great promise with the development of improved membrane which can withstand high pressure.

Advantages

- Process is simple and continuous.
- It involves no phase changes.
- The process needs extremely low energy.

(c) Electrodialysis

Principle: Electrodialysis process is based on the principle that the ions migrate towards oppositely charged electrodes when an emf is applied. The direction of ion's movement is further made specific by using ion-selective permeable membranes. The cation membranes are permeable to only cations while anion membranes are permeable to only anions. The process thus yields pure water by decreasing salt concentration.

The process of decreasing the concentration of salts in saline water using ion-selective membranes under the influence of an applied emf is called as **'electrodialysis'**.

The arrangement of an electrodialysis unit is as shown in Figure 8.7.

An electrodialysis cell consists of a series of alternative cation and anion permeable membranes (C & A). The anode is placed near the anion permeable membrane while the cathode is placed near the cation permeable membrane. A suitable emf depending on the level of salts in brine is applied

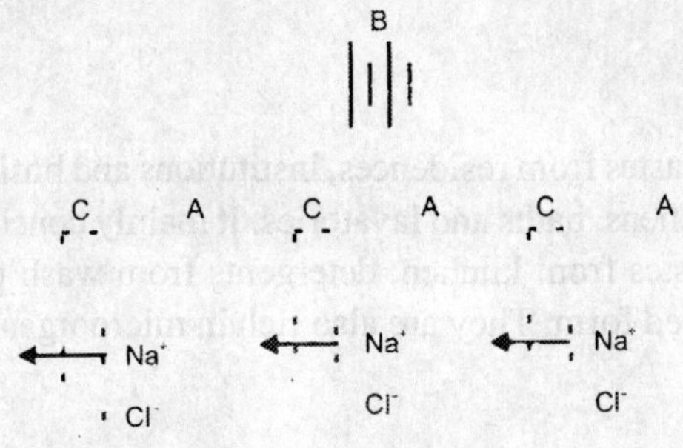

B — Source of DC supply.
A – Anion permeable membrance.
C – Cation permeable membrane.
CA – Ions losing compartment.
AC – Ions gaining compartment.

Fig. 8.7. Desalination by electrodialysis

across the two electrodes immersed in brine. Under the influence of the applied emf, sodium ions (Na^+) move through the cation permeable membrane (C) while anions (Cl^-) move through the anion permeable membrane (A) from each compartment of 'CA'. The net result is the depletion of salt content in the 'CA' compartments and an increase in salt concentration in compartments 'AC'. The fresh water produced in ion depleted compartments (CA) is collected and pumped off. The concentrated brine produced in ion concentration compartments, is discharged through the outlet.

8.7. SEWAGE AND ITS TREATMENT

Sewage is the liquid waste which includes human and house hold waste water, industrial wastes, ground wastes, street washings and storm water. Sewage contains organic and inorganic matters in dissolved, suspended and colloidal states in water. Sewage may be broadly classified into:

(i) Domestic sewage and
(ii) Industrial sewage.

(i) Domestic Sewage

Domestic sewage is the liquid wastes from residences, institutions and business buildings, containing largely organic wastes from kitchens, baths and lavatories. It mainly consists of organic load such as human excreta, urine, food wastes from kitchen, detergents from wash places and bath rooms, in suspended, dissolved or dispersed form. They are also rich in microorganisms like bacteria.

(ii) Industrial Sewage

Industrial sewage is from industrial establishments such as chemical plants, fertilizer factories, tanneries, distilleries, refineries, textiles, etc. They contain wide variety of substances, such as acids, oil, toxic chemicals, metals, animal and plant matters, non biodegradable compounds, etc.

In addition to these two types, rain water from houses, roads, etc., also constitute the sewage.

8.8. CHARACTERISTICS OF DOMESTIC SEWAGE

(a) Physical Characteristics

(i) Color and Odour: Fresh domestic sewage is, usually grey-green to grey yellow in color, but darkens with time due to decomposition. Fresh domestic sewage is without any bad smell, but when becomes stale, it develops an offensive smell due to evolution of gases like hydrogen sulfide, ammonium sulfide, phosphine, etc.

(ii) Turbidity: Sewage is normally turbid due to the presence of suspended matter.

(iii) Temperature: The temperature of sewage is slightly higher than ordinary water.

(b) Chemical Characteristics

Sewage contains more than 99% of water and remaining portion is solids in the suspended or dissolved form. It contains complex organic matter from faeces, urine, etc. These compounds can be classified as nitrogen containing and those without nitrogen. Urea, proteins, etc, are nitrogenous compound, while fats, soaps, carbohydrates are non nitrogenous compounds. Sewage also contains inorganic salts and alkalis from bathrooms, kitchens, etc. Normally fresh sewage is alkaline in nature, but becomes acidic as it becomes stale.

(c) Biological Characteristics

Sewage contains living organisms such as bacteria, algae, fungi and protozoa. Bacteria in sewage carry out the process of breaking the complex compounds into simple and stable compounds. Bacterias can be pathogenic, which are disease causing, or non-pathogenic which are harmless.

Aerobic and anaerobic oxidation: Larger number of bacterias present in sewage, are of two types, depending upon the manner in which they satisfy their oxygen demands.

(i) Aerobic bacterias: which live on free oxygen or on oxygen dissolved in water.

(ii) Anaerobic bacterias: which live and develop in the absence of free oxygen. They extract oxygen from oxygen containing radicals of organic compounds and minerals such as nitrates, nitrites and sulfates.

In the presence of oxygen, organic compounds are oxidized by aerobic bacterias and the oxidation products are nitrites, nitrates, sulfates, phosphates, etc., which do not produce any offensive smell. This kind of oxidation is known as **aerobic oxidation**.

When the oxygen content is below a certain value, the sewage is said to be stale and the organic compounds are decomposed by anaerobic bacterias, producing methane, hydrogen sulfide, phosphine, etc., which give offensive odour. This type of oxidation is called anaerobic oxidation. When the anaerobic decomposition is continuing the sewage is known as **septic sewage**.

Biological oxygen demand (BOD): When aerobic conditions prevail in the sewage, the organic impurities are decomposed by aerobic bacterias by using dissolved oxygen present in the sewage water. As a result, the dissolved oxygen content decreases. As the amount of biologically oxidisable impurities increase in water, the dissolved oxygen content decreases. Therefore, the amount of biologically oxidisable organic impurities can be expressed in terms of oxygen demand for the decomposition of these impurities.

Definition: Biological oxygen demand (BOD) of a sewage is defined as the amount of oxygen required for the biological oxidation of the organic matter under aerobic conditions at 20°C and for a period of 5 days.

Characteristics of BOD parameter

- The unit of BOD is $mg.dm^{-3}$ or ppm.
- It is empirical and semi quantitative.
- It represents only biodegradable organic load in sewage. Strictly aerobic conditions are needed.
- Determination is slow and time consuming method.

BOD indicates the amount of decomposable organic matter in the sewage. It is an expression of how much oxygen is needed for microbes to oxidize the organic matter in the sewage. It gives information about the following:

- Polluting power of sewage or its nuisance value.
- The load of organic matter on the sewage treatment plants.
- The amount of clean diluting water required for disposal of sewage.

It is important to know the BOD of sewage before disposing into rivers or lakes because, dissolved oxygen content in the water will be decreased by the sewage if its BOD is high resulting in the death of fishes and other aquatic animals.

Determination of dissolved oxygen

Dissolved oxygen content in water is determined by Winkler's method.

Principle

In this method, the given sample of water is treated with manganous sulfate and alkaline potassium iodide solution. In alkaline medium, dissolved oxygen oxidizes Mn^{2+} to Mn^{4+}, which gets precipitated as $MnO_2.H_2O$ (H_2MnO_3). On acidification, Mn^{4+} oxidizes iodide to free iodine, and the liberated iodine is then titrated with standard sodium thiosulfate solution using starch as indicator.

Fixation of dissolved oxygen,

$$Mn^{2+} + 2OH^- + \frac{1}{2} O_2 \rightarrow MnO_2.H_2O$$

Liberation of iodine,

$$H_2MnO_3 + 4H^+ + 2I^- \rightarrow Mn^{2+} + 3H_2O + I_2$$

$$I_2 + 2Na_2S_2O_3 \rightarrow Na_2S_4O_6 + 2NaI$$

Procedure

2 mL of $MnSO_4$ solution and 3 mL of alkaline potassium iodide solution are added into the bottle containing 250 mL of water sample. The bottle is stoppered and shaken well for 10 – 15 minutes, and allowed to stand for 2 minutes for the precipitate to settle. Then 1 mL of conc. H_2SO_4 is added, stoppered and shaken well to dissolve the MnO_2 precipitate. 100 mL of the solution are pipetted out from the bottle into a conical flask and titrated with standard $Na_2S_2O_3$ solution using starch as indicator till the discharge of violet colour.

Calculation

Normality of $Na_2S_2O_3$ solution ... $= N_1$

Volume of $Na_2S_2O_3$ solution consumed ... $= V_1$ mL

Normality of the dissolved oxygen $= \dfrac{V_1 \times N_1}{100}$

Weight of dissolved oxygen/ litre of water $= \dfrac{V_1 \times N_1 \times 8g}{100}$

$$= \frac{V_1 \times N_1 \times 8 \times 10^6}{100 \times 1000} \text{ ppm}$$

Determination of BOD: The parameter is commonly measured by the quantity of oxygen utilized by aerobic bacteria during 5 days period.

$$\underset{\text{Biodegradable load}}{(CHONS)} + O_2 \xrightarrow{\text{Aerobic bacteria}} CO_2 + H_2O + NO_3^- + SO_4^{2-}$$

Winkler's method for BOD measurement: In this method BOD is determined by measuring the dissolved oxygen (DO) content before and after 5 days incubation period by indirect iodometric method.

Principle: In this method, the given sample of water is treated with manganous sulfate and alkaline potassium iodide solution. In alkaline medium, dissolved oxygen oxidizes Mn^{2+} to Mn^{4+}, which gets precipitated as $MnO_2.H_2O$ (H_2MnO_3). On acidification, Mn^{4+} oxidizes iodide to free iodine, and the liberated iodine is then titrated with standard sodium thiosulfate solution using starch as indicator.

Fixation of dissolved oxygen,

$$Mn^{2+} + 2OH^- + \frac{1}{2} O_2 \rightarrow MnO_2.H_2O$$

Liberation of iodine,

$$H_2MnO_3 + 4H^+ + 2I^- \rightarrow Mn^{2+} + 3H_2O + I_2$$

$$I_2 + 2Na_2S_2O_3 \rightarrow Na_2S_4O_6 + 2NaI$$

Procedure: A definite volume of sewage sample is diluted to a known volume with dilution water which contains nutrients for bacterial growth and sufficient free oxygen. Equal volumes of diluted sample are filled into two BOD bottles.

Blank Titration: DO content in one of the bottles is determined immediately.

2ml of $MnSO_4$ solution, 3ml of alkaline KI solution are added into the bottle, stoppered, mixed well by shaking for 10–15 minutes, and allowed to stand for 2 minutes. 1ml of conc. H_2SO_4 is then added, stoppered and mixed well to dissolve the MnO_2 precipitate. A known volume of this solution is titrated with standard $Na_2S_2O_3$ solution using starch as indicator till the discharge of violet color. Repeated for agreeing values.

Sample Titration: The second bottle is incubated for 5 days at 20°C. After 5 days unconsumed DO is measured as described above.

$$BOD = \frac{D_1 - D_2}{A} \times B \text{ mg.dm}^{-3}$$

where D_1 is the DO in mg.dm^{-3} in the solution at the start. D_2 is the DO in mg.dm^{-3} in solution after 5 days. A is the volume of sample in ml before dilution and B is the volume of sample in ml after dilution.

Example 8.3: 20 ml of sewage sample was diluted to 600 ml and equal volumes were filled in two BOD bottles. DO in one bottle was determined immediately and 200 ml. of the solution required 4.2ml of N/40 $Na_2S_2O_3$ solution. The second sample was incubated for 5 days and in DO determination, 200 ml solution required 2.2 ml of N/40 $Na_2S_2O_3$ solution. Calculate BOD of the sample.

Solution:

Given,

Volume of the sample before dilution, A	= 20 ml
Volume of the sample after dilution, B	= 600 ml
Volume of the sample used for titration	= 200 ml
Volume of $Na_2S_2O_3$ solution required for blank	= 4.2 ml
Volume of $Na_2S_2O_3$ solution required for sample after 5 days	= 2.2 ml
Normality of $Na_2S_2O_3$ solution	= N/40

(a) DO of the blank solution.

$$N_1V_1 = N_2V_2$$

$$N_1 \times 200 = \frac{}{40} \times 4.2$$

$$\text{Normality of DO} = \frac{1}{200} \times \frac{1}{40} \times 4.2$$

$$\text{DO in the blank } D_1 = \frac{1}{200} \times \frac{1}{40} \times 4.2 \times \text{ eq. wt of oxygen}$$

$$= \frac{1}{200} \times \frac{1}{40} \times 4.2 \times 8$$

$$= 4.2 \times 10^{-3} \text{ g dm}^{-3}$$

$$= \mathbf{4.2 \text{ mg. dm}^{-3}}$$

(b) DO of the sample after 5 days:

$$N_1 \times 200 = \frac{1}{40} \times 2.2$$

$$\text{Normality of DO after 5 days} = \frac{1}{200} \times \frac{1}{40} \times 2.2$$

$$\text{DO in the sample, } D_2 = \frac{1}{200} \times \frac{1}{40} \times 2.2 \times 8$$

$$= 2.2 \times 10^{-3}\ \text{g.dm}^{-3}$$

$$= \mathbf{2.2\ mg.\ dm^{-3}}$$

Threrefore,

$$\text{BOD} = \frac{D_1 - D_2}{A} \times B$$

$$= \frac{4.2 - 2.2}{20} \times 600$$

$$= \mathbf{60\ mg.\ dm^{-3}}$$

Chemical oxygen demand (COD): BOD value takes into account only biologically oxidisable impurities and does not account for biologically non oxidisable and very slowly oxidisable impurities. Therefore, chemical oxygen demand (COD) parameter is introduced to measure the total oxidisable impurities present in the sewage. This includes both biologically oxidisable and biologically inert but chemically oxidisable organic matter present in the sewage.

Definition: COD is defined as the amount of oxygen used while oxidizing the total organic load of the sample with a strong chemical oxidant, $K_2Cr_2O_7$ in acid medium. It is represented in mg.dm^{-3} or ppm.

Characteristics of COD parameter

- It is a satisfactory, quantitative method for measuring total organic load..
- It is preferable to BOD as the results are reliable.
- Rapidly measurable parameter and needs about 3 hours for completion.
- In general COD > BOD since both biodegradable and non biodegradable organic load are completely oxidised.
- When used along with BOD test, it gives biologically resistant organic matter.

Determination of COD

Principle: A suitable aliquot of the sample is refluxed with a known excess of $K_2Cr_2O_7$ solution in H_2SO_4 medium and in the presence of Ag_2SO_4 and $HgSO_4$. $K_2Cr_2O_7$ oxidizes all oxidisable impurities. Ag_2SO_4 catalyses oxidation of straight chain organic compounds, aromatics and pyridine. $HgSO_4$ avoids the interference of Cl^- ions by forming soluble complex with them.

The amount of unconsumed $K_2Cr_2O_7$ is determined by titration with standard ferrous ammonium sulfate solution. The amount of $K_2Cr_2O_7$ solution consumed corresponds to the COD of the sewage sample.

Procedure: To an aliquot of sample taken in a 250 ml conical flask with ground glass joint, 1g of $HgSO_4$ and 1g of Ag_2SO_4 are added, followed by a known volume of $K_2Cr_2O_7$ solution and acidified with dilute H_2SO_4. The conical flask is fixed with a water condenser and the solution is refluxed for 2 hours. The contents are cooled and titrated with standard ferrous ammonium sulfate solution, using ferroin as indicator, till the color changes from blue green to reddish brown.

Blank Titration: The above procedure is repeated by taking the same volume of distilled water in place of the sample.

Calculations:

Volume of sample taken = V ml

Volume of std FAS used in sample titration = A ml

Volume of std FAS in Blank titration = B ml

Normality of FAS solution = N

Amount of $K_2Cr_2O_7$ consumed in satisfying the COD in term of FAS solution = (B – A) ml

$$N_1 \times V = N \times (B - A)$$

Normality of COD of the sample, $N_1 = \dfrac{N \times (B - A)}{V}$

COD of the sample $= \dfrac{N \times (B - A)}{V} \times 8 \text{ g.dm}^{-3}$

$$= \frac{N \times (B - A)}{V} \times 8 \times 1000 \text{ mg.dm}^{-3}$$

Example 8.4: 20 ml of sewage sample for COD is reacted with 25 ml of $K_2Cr_2O_7$ solution and the unreacted $K_2Cr_2O_7$ requires 9.0 ml of N/4 FAS solution. Under similar conditions, in blank titration 15.0 ml of FAS is used up. Calculate the COD of the sample.

Solution:

Given,

Volume of sample taken = 20 ml

Volume of N/4 FAS required to react with unconsumed $K_2Cr_2O_7$ = 9.0 ml

Volume of FAS consumed in blank = 15.0 ml

Amount of $K_2Cr_2O_7$ consumed to satisfy COD in terms of FAS solution = 15.0 – 9.0 = 6 ml

$$N_1V_1 = N_2V_2$$

$$N_1 \times 20 = 0.25 \times 6.0$$

Therefore, normality of COD sample, $N_1 = \frac{0.25 \times 6}{20}$

COD of the sample $= N_1 \times$ eq. wt. of oxygen

$$= \frac{0.25 \times 6}{20} \times 8$$

$$= 0.6 \text{ g. dm}^{-3}$$

$$= \mathbf{600 \text{ mg. dm}^{-3}}$$

Sewage Treatment: The domestic sewage contains heavy load of BOD, pathogenic bacteria, color and annoying smell. If such raw waste water is discharged into natural water bodies like rivers, lakes and sea, they get contaminated causing following possible effects:

- Depletion of dissolved oxygen – destruction of fish and other aquatic life.
- Color and smell affecting the quality of water.
- Pathogenic bacteria causing water borne diseases.

The domestic sewage, therefore, needs proper treatment before it is either discharged into natural water bodies or recycled. The treatment aims at the removal of oxygen demanding organic compounds, suspended and floating matter, and other matters such as ammonium salts, phosphates, pathogenic bacteria, etc. The sewage treatment is carried out in three stages.

(i) Primary treatment
(ii) Secondary treatment
(iii) Tertiary treatment

(i) Primary treatment: Primary treatment involves:
(a) Screening
(b) Silt and grit removal
(c) Oil and gas removal
(d) Sedimentation

(a) Screening: It is a physical process which removes large suspended or floating matter in the sewage. This is accomplished by using bar screens and mesh screens which retain the floating and suspended coarse particles when sewage water is passed through it.

(b) Silt and grit removal: Grit, i.e., sand, broken glass, etc., are removed by passing sewage through grit chambers, in which the velocity of flow of sewage is reduced. Being heavier, silt and grit particles settle down at the bottom.

(c) Removal of oil, grease, etc.: Oil, grease, etc., are removed in skimming tanks. They are converted into a soapy mixture by blowing compressed air through sewage water in skimming tanks, and lifted to the surface in the form of a foam. The floating substance is skimmed off.

(d) Sedimentation process: It removes finer suspended impurities. This is brought about by plain sedimentation in a continuous flow type sedimentation tank or by sedimentation with coagulation. The coagulants used are alum, ferrous sulfate, etc., which help in the easy settlement of the finely suspended particles.

(ii) Secondary treatment (Biological treatment): This process involves an aerobic biochemical oxidation or aeration. The sewage water, after sedimentation, is subjected to aerobic oxidation, during which the organic matter is converted into CO_2, the nitrogen into ammonia, and finally into nitrites and nitrates. Bases present in the sewage water form salts like ammonium nitrite, ammonium nitrate, calcium nitrate, etc.

Secondary treatment or biological treatment is generally accomplished by either trickling filter method or activated sludge process.

Trickling filter method: Trickling filter consists of a rectangular or circular vessel, with a filter bed made up of broken rock, broken bricks or large anthracite coal (Fig. 8.8). Sewage is sprayed over this bed by means of a rotating distributor. As the sewage trickles or percolates downwards through the filter bed, micro organisms grow on the surface of aggregates; using organic materials of the sewage as food. Aerobic conditions are maintained and the purified sewage is removed from the bottom. The process removes 90% of biologically oxidisable impurities.

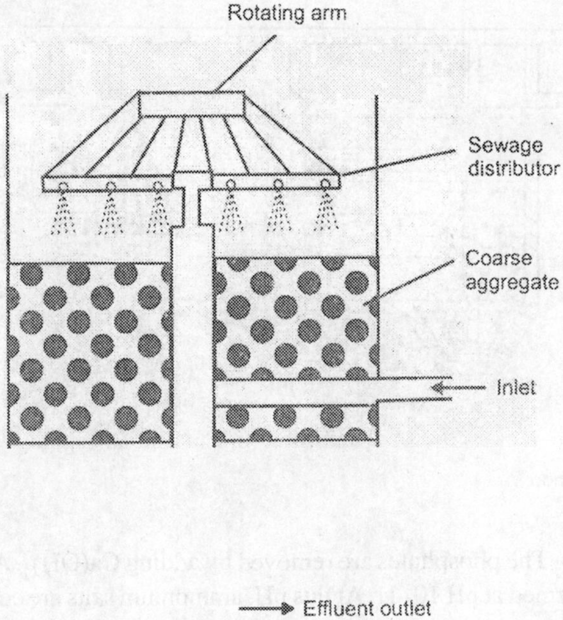

Fig. 8.8. Trickling filter

Activated sludge process: Activated sludge process involves extensive aeration of the sewage water and the process of aerobic oxidation being enhanced by the addition of a part of sludge from previous oxidation process, into sewage water. The added sludge from the previous oxidation batch is known as activated sludge, since it contains large number of aerobic bacteria and other micro organisms.

The sedimented sewage water is mixed with proper quantity of activated sludge and the mixture is sent to the aeration tank (Fig 8.9), in which the mixture is aerated and agitated for several hours. During this process, organic matters are oxidized. After the process is complete, the effluent is sent to a sedimentation tank, where sludge is deposited and water free from organic matter is drawn off. A part of the settled sludge is sent back for seeding fresh batch of sewage.

The effluent from secondary treatment has much lower organic load. This water, after chlorination to kill the pathogenic bacteria, is discharged into lakes, streams, rivers, and sea.

(iii) Tertiary treatment: The aim of tertiary treatment is further purification of waste water as well as its recycling.

The tertiary treatments consists of

(a) Removal of phosphate

(b) Coagulation and sedimentation

(c) Filtration.

(d) Degasification (NH_3 Stripping)

(e) Disinfection

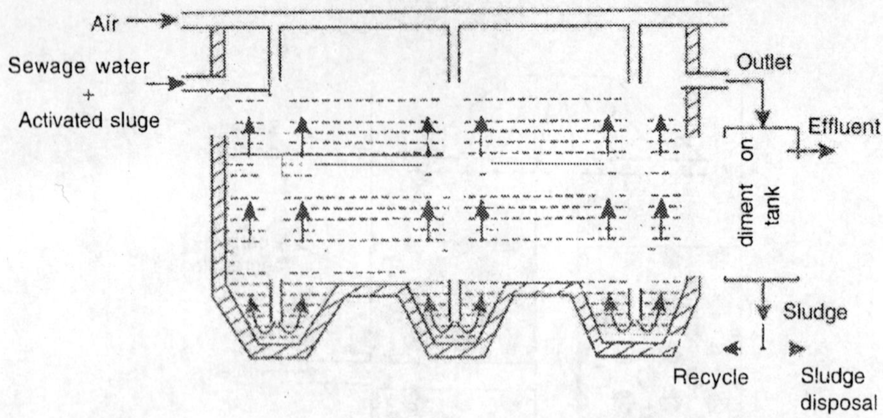

Fig. 8.9. Activated sludge method

(a) Removal of Phosphate: The phosphates are removed by adding $Ca(OH)_2$. A flocculent precipitate of calcium phosphate is formed at pH 10–11. At this pH, ammonium salts are converted into ammonia.

$$3Ca(OH)_2 + 2PO_4^{3-} \rightarrow Ca_3(PO_4)_2 + 6OH^-$$

$$Ca(OH)_2 + NH_4^+ \rightarrow Ca^{2+} + NH_3 + H_2O + OH^-$$

(b) Coagulation and Sedimentation: The suspended fine particles are removed by sedimentation in the presence of coagulants like alum, ferrous sulfate, etc. The flocculent precipitates of $Al(OH)_3$ or $Fe(OH)_2$ formed by the coagulants entrap the fine particles and help them to settle down. The highly charged ions of the coagulants also neutralize the charges on colloidal particles and make them to coagulate and settle down.

(c) Filtration: For filtration, the water is passed through a conventional sand filter beds to remove the last traces of suspended matter.

(d) Degasification: Stripping of ammonia and other gases is done in a degasifier. The degasifier consists of a large tower fitted with a number of perforated plates. The hot water trickles through these plates. Large surface area and higher temperature promote stripping of dissolved gases like NH_3, CO_2, H_2S, etc.

(e) Disinfection: The pathogenic bacterias are destroyed by disinfection. Among many disinfectants, chlorine is cheap and effective.

$$Cl_2 + H_2O \rightarrow HOCl + H^+ + Cl^- \text{ (pH 6.5)}$$

Unionized HOCl attacks the cells of bacteria and kills them. The final composition of tertiary treated waste water is.

$$BOD < 1 \text{ ppm}$$

$$NH_4^+ < 1 \text{ ppm}$$

$$PO_4^{3-} < 1 \text{ ppm}$$

The treated water has high clarity, free from odour, low BOD, and therefore, it is nearly equivalent to drinking water and can be recycled.

Sludge disposal: Sludge, which is collected from sewage treatment processes are disposed off by the following methods.

(i) Burrial at sea: At places, which are near the sea, sludge can be dumped into sea.

(ii) Land spreading: The sludge is uniformly spread over soil, followed by ploughing. It acts as a fertilizer.

(iii) Septic tank treatment and sludge digestion: The sludge is kept in a closed tank in the absence of air for prolonged period (about 30 days). The sludge undergoes anaerobic decomposition producing gases like methane, H_2S, phosphine, etc. The gas can be used as fuel for city supply or power generation.

REVIEW QUESTIONS

1. What are the sources of water?
2. What are the impurities present in water?
3. What is hard water? What are the causes for hardness?
4. Differentiate between temporary hardness and permanent hardness.
5. What are scales and sledges? How are they formed?
6. What are the harmful effects of scales?
7. Explain priming and foaming. What are their causes and harmful effects? How are they prevented?
8. Give an account of boiler corrosion.
9. Explain demineralization of water.
10. Explain the following methods of desalination of water:
    (a) Multistage flash evaporation
    (b) Electrodialysis
    (c) Reverse osmosis
11. What is a sewage? How is it classified?
12. Explain the characteristics of sewage.
13. How is dissolved oxygen content in water determined?
14. Define BOD of a sewage. What are its significances? How is it determined?
15. Define COD of a sewage. How is it different from BOD? How is it determined?
16. Why does the sewage need treatment? Explain the primary, secondary and tertiary treatment methods for sewage.
17. Justify the following statements:
    (i) Stale sewage produces foul smell.
    (ii) COD is greater than BOD.

(iii) Disinfection is necessary before discharging treated sewage.
(iv) Chloramine is a better disinfectant compared to chlorine and bleaching powder.
(v) Sedimentation with coagulation is faster than plain sedimentation.
(vi) Presence of small amounts of $MgCl_2$ in boiler feed water leads to continued corrosion of boiler metal.

18. 0.5 g of $CaCO_3$ was dissolved in dil HCl and made up to 250 ml. 25 ml of this solution required 20.5 ml of EDTA solution for titration. 25 ml of the hard water sample consumed 18 ml of EDTA. 25 ml of the same water sample after boiling and filtering, required 12 ml of EDTA for titration. Calculate total hardness, temporary hardness and permanent hardness.

CHAPTER 9

Nanomaterials

9.1 INTRODUCTION

What cannot be achieved with the big and bulky, may more easily be done with the small and neat. That is why nanoscience which is the science of the tiny, has gained lot of interest during the last decades. It is a broad field connected to both chemistry, physics, medicine, material science as well as to applied technologies. The amount of possible applications is huge, and the potential benefit to technology and society are considered to be vast. Even though some applications built on concepts of nanoscience and nanotechnology have been at the market for some time, the field can still be considered a young one, and a lot of things are to be done, discovered, described, explained and exploited.

Nanomaterials are usually considered to be materials with at least one external dimension less than 100 nanometers. They may be in the form of particles, tubes, rods or fibers. The nanomaterials and bulk materials of the same composition have different physico-chemical properties. Nano-sized particles exist in nature and can be created from a variety of products, such as carbon or minerals like silver. Nanomaterials can be metals, ceramics, polymeric materials, or composite materials. The term nano comes from a Greek word meaning **dwarf** or extremely small. Most nanoscale materials are too small to be seen with the naked eye and even with conventional lab microscopes. Materials engineered to such a small scale are often referred to as **engineered nanomaterials** (ENMs), which exhibit unique optical, magnetic, electrical, and other properties. Nano-sized materials currently are used in numerous industries. Examples include:

- Carbon black particles make rubber tires wear resistant.
- Nanofibers are used for insulation and reinforcement of composites.
- Nano-iron oxide creates the magnetic material used in disk drives and audio-video tapes.
- Nano-zinc oxides and titania are used as sunblocks for UV rays.

Nanoscale particles and nanothin layers of materials are being used, among other things, to make products lighter, stronger or more conductive. Some of the products on the market using nanotechnology are:

- Magnetic recording tapes;

- Computer hard-drivers;
- Bumpers on cars;
- Solid-state compasses;
- Protective and glare-reducing coatings for eyeglasses and windows;
- Automobile catalytic converters;
- Metal-cutting tools;
- Dental bonding agents;
- Longer-lasting tennis ball;
- Burn and wound dressing;
- Ink; etc.

The emergent properties of nanomaterials have the potential for great impacts in electronics, medicine, and other fields. For example:

- Nanotechnology can be used to design pharmaceuticals that can target specific organs or cells in the body such as cancer cells, and enhance the effectiveness of therapy. One of the attractive applications in nanomedicine is the creation of nanoscale devices for improved therapy and diagnostics. Such nanoscale devices or nanorobots serve as vehicles for delivery of therapeutic agents, detectors or guardians against early disease and perhaps repair of metabolic or genetic defects.
- Nanomaterials can also be added to cement, cloth and other materials to make them stronger and yet lighter.
- Their size makes them extremely useful in electronics, and they can also be used in environmental remediation or clean-up to bind with and neutralize toxins.

The variety of nanomaterials is great, and their range of properties and possible applications appear to be enormous, from extraordinarily tiny electronic devices, including miniature batteries, to biomedical uses, and as packaging films, super absorbents, components of armor, and parts of automobiles.

9.2. DEFINITIONS

Particle

A particle is defined as a small object that behaves as a whole unit in terms of its transport and properties.

Nanomaterial

A nanomaterial is defined as that having at least one of its dimensions between 1 and 100 nm.

Nanoscience

Naboscience primarily deals with synthesis, characterization, exploration, and exploitation of nanostructured materials.

Nanotechnology

Nanotechnology is defined as the design, characterization, production and application of structures, devices and systems by controlling shape and size at the nanometer scale.

Nanotechnology integrates chemistry, physics and biology of nanostructures with materials science, electronics, and process technologies in the fabrication and manufacturing of nanodevices, nanomachines, ultra-low integrated circuits, micro optoelectro-mechanical systems, nanobiorobots, etc.

9.3. CLASSIFICATION

Based on the dimensionality, nanomaterials are classified as

- Zero dimensional (0D)
- One dimensional (1D)
- Two dimensional (2D) and
- Three dimensional (3D).

(a) Zero-Dimensional Nanomaterials

Materials having all the dimensions within the nanoscale (no dimensions, or 0-D, are larger than 100 nm). The most common examples of zero-dimensional nanomaterials are nanoparticles (Fig. 9.1a).

(b) One-Dimensional Nanomaterials

Materials having one of their dimensions outside the nanoscale (two dimensions are in the nanoscale). This leads to needle like-shaped nanomaterials. Examples include nanotubes, nanorods, and nanowires (Fig. 9.1b).

(c) Two-Dimensional Nanomaterials

Materials having two of their dimensions outside the nanoscale (only one dimension is in the nanoscale). 2D nanomaterials exhibit plate-like shapes. Examples include nanofilms, nanolayers, and nanocoatings (Fig. 9.1c).

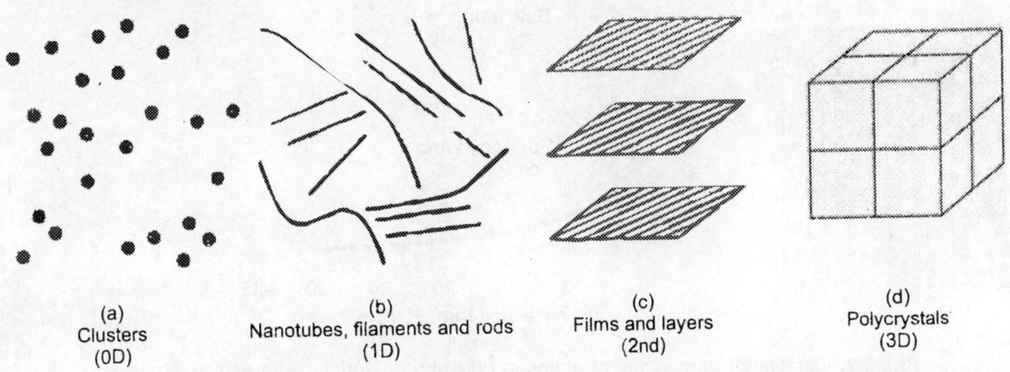

Fig. 9.1. Classification of nanomaterials (a) 0D, (b) 1D, (c) 2D and (d) 3D nanomaterials.

(d) Three-Dimensional Nanomaterials

Materials having all their three dimensions outside the nanoscale. Materials that are not confined to the nanoscale in any dimension are considered nanomaterials as these materials are composed of a multiple arrangement of nanosize crystals, most typically in different orientations. They possess a nanocrystalline structure or involve the presence of features at the nanoscale. They can contain dispersions of nanoparticles, bundles of nanowires, and nanotubes as well as multinanolayers.

9.4 PROPERTIES OF NANOMATERIALS

Nanomaterials have the structural features in between of those of atoms or molecules and the bulk materials. The properties of nanomaterials are significantly different from those of atoms or molecules and bulk materials. A bulk material possesses constant physical properties regardless of its size, but at the nanoscale it is not often the case. They exhibit size dependent properties. This is mainly due to the nanometer size of the materials which render them:

- large fraction of surface atoms to the bulk atoms,
- high surface energy,
- spatial confinement and
- reduced imperfections.

The properties of materials change as their size approaches the nanoscale and as the percentage of atoms at the surface of a material becomes significant. In bulk materials the percentage of atoms at the surface is very small relative to the total number of atoms of the material. Hence for bulk or macro materials the properties are often determined entirely by the bulk, due to the relatively small contribution by a small surface area. Due to their small dimensions, nanomaterials have extremely large surface area to volume ratio, resulting in more "surface" dependent material properties. Figure 9.2 shows the variation of percentage of bulk atoms and surface atoms with change in size. Increase

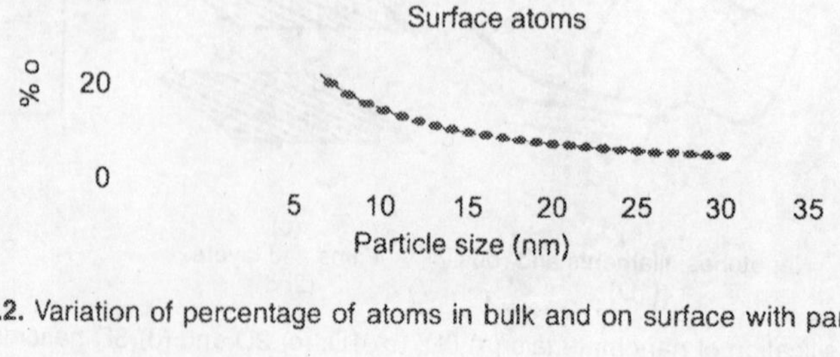

Fig. 9.2. Variation of percentage of atoms in bulk and on surface with particle size.

in the surface area to volume ratio also results in increase in the surface energy per unit quantity of the material. The interesting and sometimes unexpected properties of nanomaterials are partly due to the aspects of the surface of the material dominating the properties in lieu of the bulk properties.

The nanometer feature sizes of nanomaterials also have spatial confinement effect on the materials. For example, the delocalized electron cloud on the bulk metal remains confined within the influence of a few atoms in a nanoparticle. The imperfections such as point defects, line defects and area defects generally extended over the length and area of a bulk material gets reduced when it is converted into a nanomaterial as the size and correspondingly the number of atoms present in the nanomaterial is small. As a result, surface properties such as energy levels, electronic structure, and reactivity can be quite different from interior states, and give rise to quite different material properties.

(a) Thermal Properties

Nanomaterials may have a significantly lower melting point or phase transition temperature due to a huge fraction of surface atoms in the total amount of atoms. Atoms or molecules on a solid surface possess fewer nearest neighbors or coordination numbers and number of bonds formed by each of the surface atoms is less than that formed by the atoms in the bulk. Therefore, the number of bonds need to be broken per atom during melting is less in nanomaterial than in macromaterial, in which the number of bonds need to be broken per atom is large due to large proportion of atoms in the bulk. Fig. 9.3 shows the effect of particle size on melting point.

Thermal transport: Heat is transported in materials by two different mechanisms; by lattice vibration waves (phonons) and free electrons. In the case of bulk solid nanomaterials, the wavelengths of phonons are much smaller than the length scale of the microstructure. However, in nanomaterials, the length scale of the structure is similar to the wavelength of phonons. Therefore, quantum confinement occurs. To understand the effect one can apply the simple relation between the thermal

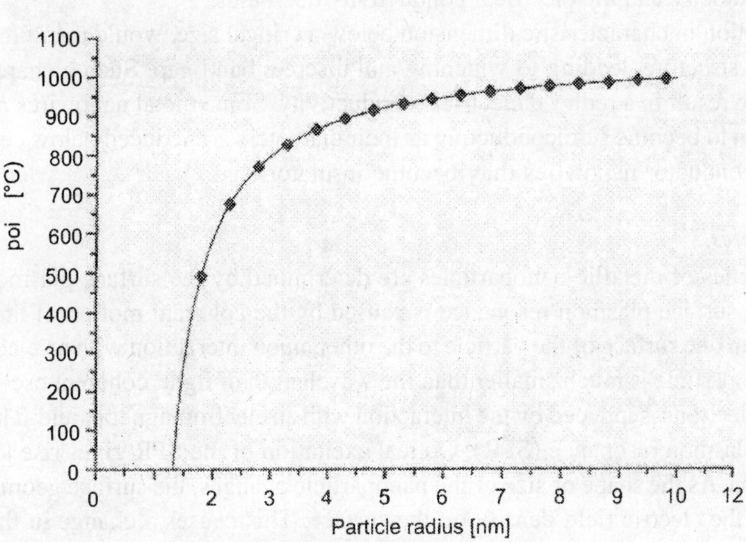

Fig. 9.3. Effect of particle size on melting point

conductivity (K) of gases with path length (l) as per kinetic theory of gases, to the thermal phonon gases:

$$K = \frac{1}{3} Cvl$$

where v is the particle velocity and C is the heat capacity of unit volume. For bulk solid of big size $d > l_{phonon}$, the size effect does not arise. However, under reducing of the size d it can become smaller than free path length $d < l_{phonon}$, resulting to a decreasing of K.

(b) Electrical Properties

In the case of bulk materials, the conduction of electrons is delocalized and electrons can move in all dimensions. As they travel their paths, the electrons are primarily scattered by various interactions with phonons, impurities and interfaces, resembling a random walk process. Reduction in material's dimensions would have the following effects on electrical resistivity.

- An increase in crystal perfection or reduction of defects, which would result in a reduction in defect scattering and, thus, a reduction in resistivity or increase in electrical conductivity. However, the defect scattering makes a minor contribution to the total electrical resistivity of metals at room temperature, and thus the reduction of defects has a very small influence on the electrical conductivity.
- An additional contribution to the total resistivity due to surface scattering, plays a very important role in determining the total electrical resistivity of nanosized materials. When the scattering is totally inelastic or diffuse, the electron mean free path is terminated by impinging on the surface. After the collision, the electron trajectory is independent of the impingement direction and the subsequent scattering angle is random. Consequently, the scattered electron loses its velocity and the electrical conductivity decreases.
- The reduction in characteristic dimension below a critical size, would result in a change of electronic structure, leading to widening and discrete band gap. Such a change generally would also result in a reduced electrical conductivity. Some metal nanowires may undergo a transition to become semiconducting as their diameters are reduced below certain values, and semiconductor nanowires may become insulators.

(c) Optical properties

The optical properties of metallic nanoparticles are determined by the surface plasmon resonance phenomenon. The surface plasmon resonance is caused by the coherent motion of the conduction band electrons from one surface of the particle to the other, upon interaction with an electromagnetic field. When a nanoparticle is much smaller than the wavelength of light, coherent oscillation of the conduction band electrons is induced by the interaction with an electromagnetic field. This resonance is called surface plasmon resonance (SPR). Optical excitation of the SPR gives rise to the surface plasmon absorption. As the shape or size of the nanoparticle changes, the surface geometry changes causing a shift in the electric field density on the surface. This causes a change in the oscillation frequency of the electrons, generating different optical properties including absorption and scattering.

In small nanoclusters the effect of reduced dimensionality has the most profound effect on the energies of highest occupied molecular orbital (HOMO) which is valence band and the lowest unoccupied molecular orbital (LUMO), essentially the conduction band. The optical emission and adsorption occurs when the transition of the electrons occurs between these two states. Semiconductors and many metals show large changes in optical properties such as color, as a function of particle size. For example colloidal suspensions of gold nanoparticles have a deep red color which becomes progressively more yellow as the particle size increases:

- Gold spheres of >20 nm exhibit purple color,
- Gold spheres of 10-20 nm exhibit red color,
- Gold spheres of 2-5 nm exhibit yellow color.

The energy gap between the valence band and the conduction band (corresponding to the HOMO-LUMO gap in molecular compounds) widens as the particle size decreases. The corresponding absorption band is blue-shifted and becomes sharper. Other properties which may be affected by reduced dimensionality include photocatalysis, photoconductivity, photoemission and electroluminescence.

(d) Magnetic Properties

Magnetic properties of nanostructured materials are distinctly different from that of bulk materials. The large surface area to volume ratio results in a substantial proportion of atoms having different magnetic coupling with neighboring atoms leading to differing magnetic properties. Bulk gold and platinum are non-magnetic but at the nano size they act as magnetic particles. Au nanoparticles become ferromagnetic when they are capped with the appropriate molecules such as thiol.

Ferromagnetism disappears and transfers to superparamagnetism in the nanometer scale due to the huge surface energy. Magnetic nanoparticles are used in a range of applications like imaging, bioprocessing, refrigeration as well as high storage density magnetic memory media. Giant magnetoresistance (GMR) is a phenomenon observed in nanoscale multilayers consisting of strong ferromagnet (Fe, Co, Ni) and a weaker magnetic or non-magnetic buffer (Cr, Cu). It is usually employed in data storage and sensing.

(e) Mechanical Properties

Mechanical properties of nanomaterials may reach the theoretical strength, which are one or two orders of magnitude higher than that of the bulk materials. For example, the bending of bulk copper (wire, ribbon, etc.) occurs readily with the movement of copper atoms / clusters at about 50 nm scale. Copper nanoparticles smaller than 50 nm are considered super hard materials that do not exhibit the same malleability and ductility as bulk copper.

The enhancement in mechanical strength is simply due to the reduced probability of defects. Imperfections in bulk materials, such as dislocations are often created to accommodate stresses generated in the synthesis and processing of bulk materials due to temperature gradient and other inhomogeneities. The smaller the cross-section of a material, the less is the probability of finding in it any imperfections such as dislocations, micro-twins, impurity precipitates, etc.

9.5. SYNTHESIS OF NANOMATERIALS

There are two general approaches to the synthesis of nanomaterials and the fabrication of nanostructures:

- The bottom-up approach, that is the miniaturization of the components.
- The top-down approach, that involves the self-assembly of molecular components.

Attrition or milling is a typical top-down method in making nanoparticles, whereas the colloidal dispersion is a good example of bottom-up approach in the synthesis of nanoparticles. Bottom-up approaches seek to have smaller components arrange themselves into more complex assemblies, while top-down approaches seek to create nanoscale devices by using larger, externally-controlled ones to direct their assembly. In the bottom-up approach, the nanomaterial is built by assembling it from building blocks (such as atoms and molecules), atom-by-atom, molecule-by-molecule, or cluster-by-cluster. In top-down approach, the preparation is started with the bulk material and is cut away to get the nanomaterial. The schematic representation of the building up of nanostructures is represented as follows:

Bottom up approach plays an important role in the fabrication and processing of nanomaterials for the following reasons:

- When the structures fall into the nanometer scale, there is little choice for a top-down approach as all the tools possessed are too big to deal with such tiny objects.

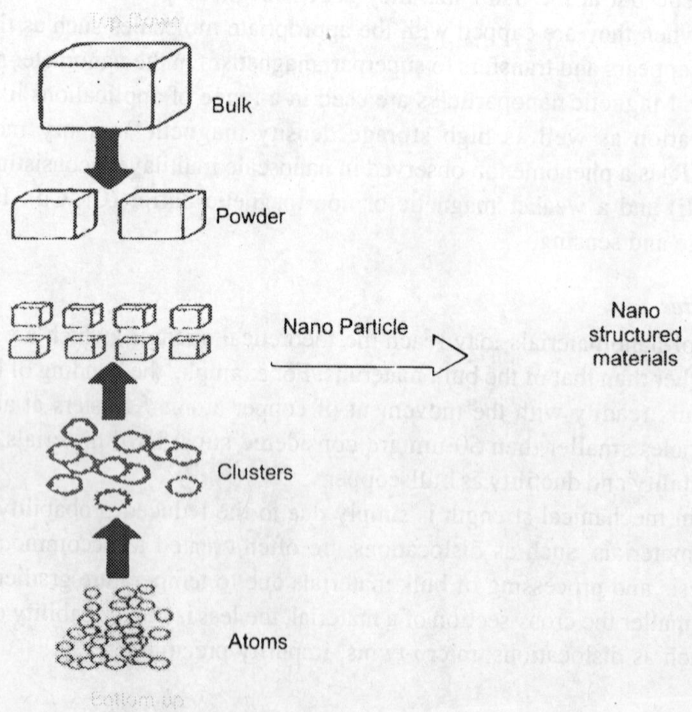

Schematic representation of the building up of nanostructures.

- Bottom-up approach promises a better chance to obtain nanostructures with less defects, more homogeneous chemical composition.
- Top-down approach likely to introduce internal stress. in addition to surface defects and contaminations.

Nanoparticles can be synthesized in many different ways, both through chemical reactions and physical processes. Most common methods used for the commercial or industrial manufacture of nanoparticles may be divided into four main groups:

- ***Gas phase processes*** including vapor deposition, flame pyrolysis, high temperature evaporation and plasma synthesis.
- ***Liquid phase methods*** in which chemical reactions in solvents lead to the formation nanoparticles.
- ***Sol-gel technique.***
- ***Solid phase mechanical processes*** including grinding, milling and alloying.

Even for the same material different methods are often used in order to optimize specific properties of nanoparticles such as size, size distribution, symmetry, purity and others.

(a) Sol-gel Process

Sol-gel process offers possibility to produce nanomaterials with novel, predefined properties in a simple process and at relatively low process cost. The sol is a colloidal solution made of solid particles dispersed in a liquid phase. The gel is colloidal solution made of liquid particles dispersed in a solid phase. So, in general terms, the sol-gel process consists in the chemical transformation of a sol into a gel state and with subsequent post-treatment and transition into solid nanomaterial. The method involves a set of chemical reactions which irreversibly convert a homogeneous solution of molecular reactant precursors (a sol) into a three-dimensional polymer (a gel) forming an elastic solid filling the same volume as the solution. The sol-gel process can be characterized by a series of distinct steps.

1. Preparation of precursor solution: Precursor can be inorganic, but more often metal organic precursors are used. Common metal organic precursors for the sol-gel process are metal alkoxides (MOR), where R stands for an alkyl group. In a typical sol-gel process, the precursor is subjected to a series of hydrolysis and polymerization reactions to form a colloidal suspension (the sol). Metal alkoxide is dissolved in alcohol and then water is added under acidic, neutral or basic conditions. Addition of water leads to hydrolysis in which alkoxide ligand is replaced with a hydroxyl liquid:

$$MOR + H_2O \rightarrow MOH + ROH$$

2. Formation of gel phase: The polycondensation reaction between MOH and MOR results in the formation of an oxide- or alcohol-bridged network (gel). Gel formation results in a dramatic increase in the viscosity of the solution.

$$MOH + MOR \rightarrow M\text{-}O\text{-}M + ROH$$

3. Aging of the gel (syneresis): The reaction mixture is allowed to continue polycondensation reactions until the gel transforms into a solid mass, accompanied by contraction of the gel network and expulsion of solvent from gel pores. Ostwald ripening (is the phenomenon by which smaller particles are consumed by larger particles during the growth process) and phase transformations may occur concurrently with syneresis.

4. Drying of the gel: Water and other volatile liquids are removed from the gel network. If isolated by thermal evaporation, the resulting nanomaterial is termed a **xerogel.** If the solvent is extracted under supercritical or near supercritical conditions, the product is an **aerogel.**

5. Dehydration: During dehydration, surface- bound M-OH groups are removed, thereby stabilizing the gel against rehydration. This is normally achieved by calcining at temperatures up to 800°C.

6. Densification and decomposition of the gels at high temperatures (T >800°C). The pores of the gel network are collapsed, and remaining organic contaminants are volatilized.

The typical steps that are involved in sol-gel processing are shown in Fig. 9.4.

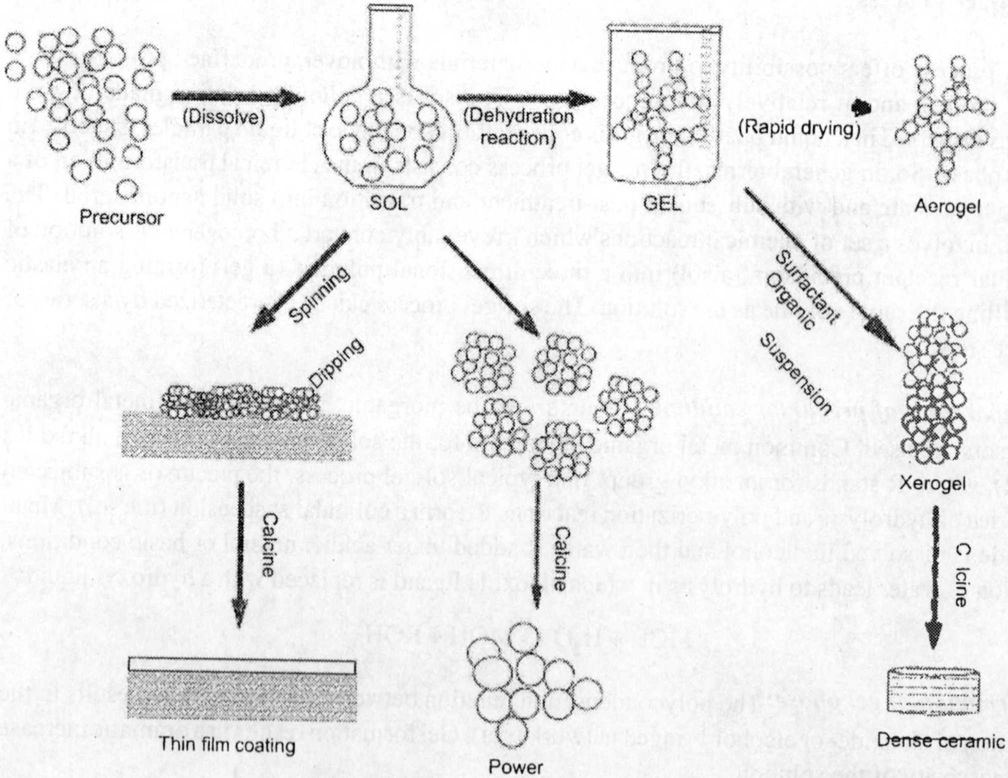

Fig. 9.4. Schematic representation of sol-gel process of synthesis of nanomaterials

Advantages:

- Excellent stoichiometry control of precursor solutions,
- Ease of compositional modifications,
- Customizable microstructure,
- Ease of introducing various functional groups or encapsulating sensing elements,
- Relatively low annealing temperatures,
- The possibility of coating deposition on large-area substrates,
- Simple and inexpensive equipment.

(b) Precipitation Process

This method relies on the precipitation of nanometer-sized particles within a continuous fluid solvent. An inorganic metal salt, such as chloride, nitrate, sulphate, etc., is dissolved in water. Metal cations exist in the solution in the form of metal hydrate species, for example, $Al(H_2O)^{3+}$ or $Fe(H_2O)_6^{3+}$. These hydrates are added with basic solutions, such as NaOH or NH_4OH. The precipitated hydrolyzed species is filtered, washed, dried and calcined in order to obtain the final nanomaterial product. Several other precipitation agents such as oxalate, carbonate, etc., can also be used.

- The method is relatively simple and widely used for production of single and multi-components oxide nanopowders using optimized reactions and reaction conditions.
- Subsequent processing of colloids can include additional colloidal precipitation on particle surfaces to produce a core-shell nanoparticle structure, deposition on substrates

(c) Gas Condensation Process

The gas condensation process is one of the most known and well established procedures for the production of nanopowders. Gas condensation, as a technique for producing nanoparticles, refers to the formation of nanoparticles in the gas phase, by the condensation of atoms and molecules in the vapor phase. Figure 9.5 gives a schematic overview of the gas condensation procedure.

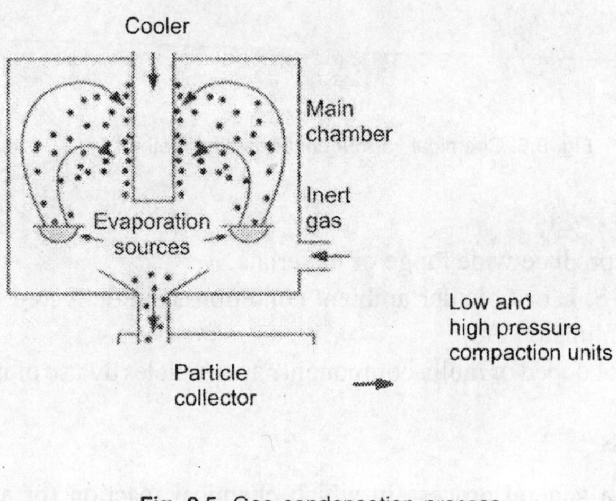

Fig. 9.5. Gas condensation process

In this method, often a metal, is evaporated from a heated metallic source into a chamber which has been previously evacuated and filled with inert gas to a low-pressure. The metal vapor cools through collisions with the inert gas atoms and condenses into nanoparticles. The particle size is usually in the range 1–100 nm and can be controlled by varying the inert gas pressure. Ultimately, the particles are collected and may be compacted to produce a dense nanomaterial.

(d) Chemical Vapor Condensation Process

In this method a mixture of gas reactants are delivered into a reaction chamber. Inside the chamber, the chemical reactions among the gas molecules are induced by an input of energy such as resistant heating, laser, and plasma. The product formed in the vapor state on condensation, deposits nanoparticles. The by-products of the reactions are exhausted. A schematic representation of the process is given in Fig. 9.6.

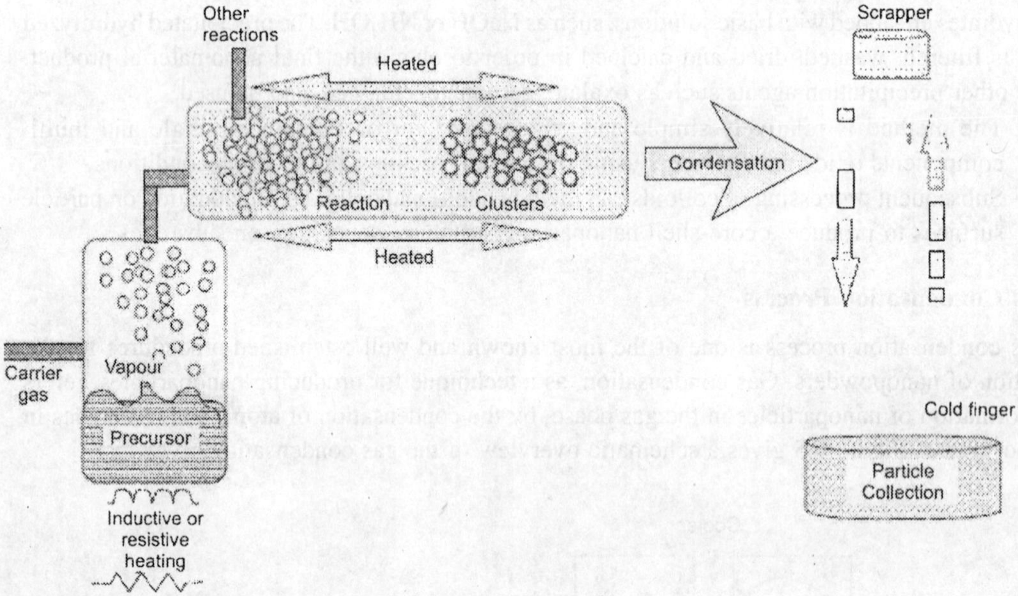

Fig. 9.6. Chemical vapour condensation method

Advantages

- Very flexible, can produce wide range of materials.
- Precursors can be S, L or G under ambient conditions; but delivered to reactor as vapor (using bubbler, sublimator, etc).
- Allows formation of doped or multi-component nanoparticles by use of multiple precursors.

(e) Hydrothermal Process

Solvothermal process is a general process in which chemical reaction (or a transformation) is carried out between precursor(s) in a solvent (in a close system) at a temperature higher than the boiling temperature of this solvent and under high pressure. **Hydrothermal process** is a special case when water is used as the solvent.

The reactants are dissolved (or placed) in water in a closed vessel, called a hydrothermal bomb or autoclave. The autoclave is heated above the boiling point of water under high pressure so that the reaction takes place in supercritical water. The temperature in maintained between 150°C and 500 °C; and the pressure between 100 kbar and 3000 kbar. A schematic diagram of the hydrothermal synthesis setup is shown in Fig. 9.7.

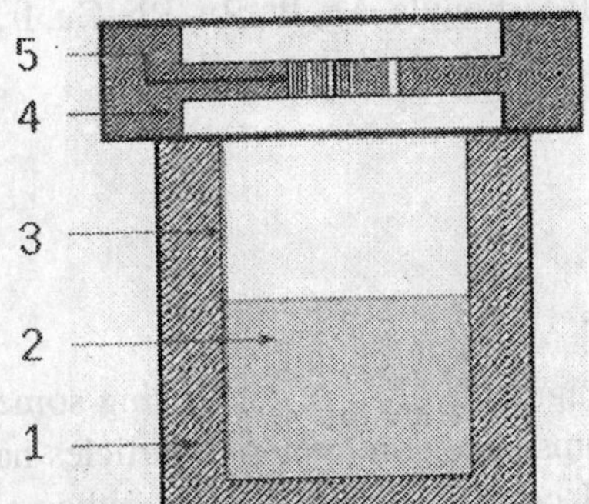

Fig. 9.7. Schematic diagram of hydrothermal synthesis setup: (1) stainless steel autoclave (2) precursor solution (3) Teflon liner (4) stainless steel lid (5) spring.

The method is used in the synthesis of

- Nano oxides (TiO_2, VO_x, M_xMnO_2),
- Layered oxides: nanowires/tubes,
- Carbon-nanotubes, and
- Some elemental nanostructures (e.g., Bi)

Substantial enhancement of the hydrothermal method facilitates the use of additional external factors to control the reaction medium during the synthesis process. The approach is implemented in the hydrothermal-microwave, hydrothermal-ultrasonic, hydrothermal-electrochemical and hydrothermal-mechanochemical synthesis methods.

Advantages

- Low cost.
- Environmentally friendly technique that can be used on large area as well as fabrication of many nanostructures.
- The low substrate temperature enables this method to create various nanostructures on temperature sensitive substrates.

(f) Thermolysis Process

The natural synthesis of nanoparticles by "thermolysis" is characterized by subjecting the metal precursors (usually organometallic compounds in oxidation state zero) at high temperatures together with a stabilizing compound to thermal decomposition. Thermal decomposition provides remarkable control over size and is well suited for scale up to larger quantities.

Various metal nanoparticles have been prepared by the thermal decomposition of low-valent complexes involving olefinic an carbonyl ligands. For example cobalt nanoparticles are prepared by the thermal decomposition of cobalt carbonyls. Ag, Pt, Ru, Rh, Cu, Ir, Zn, Co and Ni nanoparticles are prepared by the decomposition of their olefinic complexes.

9.6. NANOSCALE MATERIALS

(a) Nanocrystals or Nanoclusters

Nanoclusters or nanocrystals are fragments of solid comprising somewhere between a few atoms and a few tens of thousands of atoms. They are material particles having at least one dimension smaller than 100 nanometers and composed of atoms in either a single- or poly-crystalline arrangement. They are made up of equal subunits, such as, atoms of a single element, molecules or even combinations of atoms of several elements in subunits with equal stoichiometries (alloys, etc.).

Examples: $(Na)_n$, $(SF_6)_n$, $(H_2O)_n$, $(Cu_3Au)_n$, $(ClCH_3C_6H_3CO_2H)_n$, $(TiO_2)_n$, etc.

These can include, e.g., fullerenes, metal clusters (agglomerates of metal atoms), large molecules, such as proteins, and even hydrogen-bonded assemblies of water molecules, which exist in water at ambient temperatures.

(b) Fullerenes

Fullerenes are made by heating of graphite in an electric arc in the presence of inert gases such as helium or argon. The sooty material formed by the condensation of carbon vapor consists of mainly C_{60} with smaller quantity of C_{70} and traces of fullerenes consisting of even number of carbon atoms up to 350 or above. Fullerenes are cage-like molecules. C_{60} molecule has a shape like soccer ball and called **Buckminsterfullerene** (Fig. 9.8).

It contains twenty-six-membered rings and twelve five- membered rings. A six-membered ring is fused with six- or five-membered rings but a five-membered ring can only fuse with six-membered rings. All the carbon atoms are equal and they are sp^2 hybridized. Each carbon atom forms three sigma bonds with other three carbon atoms. The remaining electron at each carbon is delocalized in molecular orbitals, which, in turn, give aromatic character to molecule. This ball shaped molecule has 60 vertices and each one is occupied by one carbon atom and it also contains both single and double bonds with C–C distances of 143.5 pm and 138.3 pm respectively. Spherical fullerenes are also called Bucky balls in short.

The fullerenes are soluble in common solvents such as benzene, toluene or chloroform.

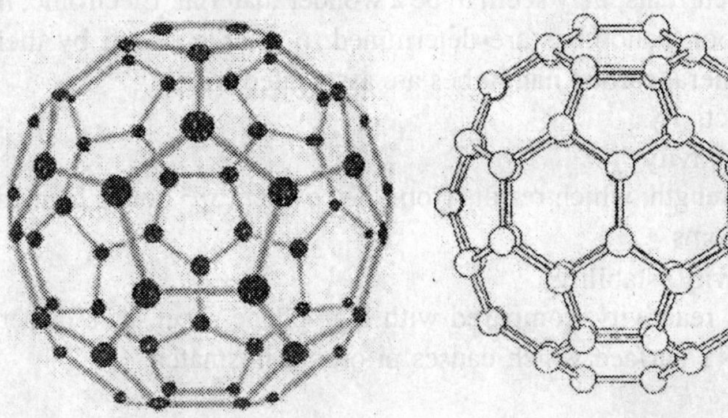

Fig. 9.8 Structure of C_{60}

Uses

- In optics the fullerenes were proposed to use for the electrophotographical imaging, optical filters, photopolymer sensibilizators, nonlinear optical materials.
- In electronics they are used as the molecular switchers, diodes, transistors, resistors for lithography, collar cell elements, magneto-optical recorders, photoelectronic devices.
- In military sphere the fullerenes are used as the optical and microwave absorption coatings to prevent aircraft and missiles from radar detection.
- Fullerenes are chemically reactive and can be added to polymers and elastomers to create new copolymers with specific physical and mechanical properties.

(c) Carbon nanotubes

Nanotubes are the 1D quantum cylinders nanoscopic in diameter while micro/macroscopic in length so they are quantum macrocrystals possessing potential unique properties. Carbon nanotubes are members of the fullerene structural family. Their name is derived from their long, hollow structure with the walls formed by one-atom-thick sheets of carbon, called graphene. Carbon nanotubes are single sheets of graphene rolled into cylinders. The diameter of the tubes are typically of nanometer dimensions, while the lengths are typically micrometers. Carbon nanotubes are categorized as single-walled nanotubes (SWNTs) and multi-walled nanotubes (MWNTs). Individual nanotubes naturally align themselves into "ropes" held together by van der Waals forces. Multi-walled nanotubes (MWNT) consist of multiple rolled layers (concentric tubes) of graphene.

Carbon nanotubes are generally produced by three main techniques, arc discharge, laser ablation and chemical vapor deposition. In arc discharge, a vapor is created by an arc discharge between two carbon electrodes with or without catalyst. Nanotubes self-assemble from the resulting carbon vapor. In the laser ablation technique, a high-power laser beam impinges on a volume of carbon containing feedstock gas (methane or carbon monoxide).

Properties: Carbon nanotubes, with one hundred times the tensile strength of steel, thermal conductivity better than all materials, and electrical conductivity similar to copper, but with the ability to carry much higher currents, they seem to be a wonder material. Electronic, molecular and structural properties of carbon nanotubes are determined to a large extent by their nearly one dimensional structure. In general carbon nanotubes are associated with:

- High electric conductivity.
- High thermal conductivity .
- High mechanical strength, which results from the covalent sp^2 bonds formed between the individual carbon atoms.
- High thermal resistivity / stability.
- Enhanced chemical reactivity, compared with a graphene sheet, as a direct result of the curvature of the CNT surface which causes pi-orbital mismatch.

Uses

- Field Emitters/Emission: Conductive or reinforced plastics
- Molecular electronics: CNT based non-volatile RAM, CNT based transistors
- Energy storage
- CNT based fibers and fabrics
- CNT based ceramics
- Biomedical applications

(d) Nanowires

A **nanowire** is a nanostructure, with the diameter of the order of a nanometer with the ratio of the length to width being greater than 1000. Alternatively, nanowires can be defined as structures that have a thickness or diameter constrained to tens of nanometers or less and an unlimited length. They are also referred to as quantum wires. Nanowires can be metallic (Ni, Pt, Au, etc.), semiconducting (Si, GaN, etc.) or insulating (SiO_2, TiO_2, etc.). They belong to one-dimensional (1-D) nanomaterials.

Nanowires are prepared by using a number of techniques, including suspension, electrochemical deposition, vapor deposition, etc. But the most commonly used technique is vapor-liquid-solid (VLS) synthesis. The method uses either laser ablated particles or a feed gas as the source materials. Liquid metal (Au) nanoclusters in the form of a thin film on a substrate is used as the catalyst. The source enters the nanoclusters and begins to saturate them. On attaining super saturation, the source solidifies and grows outward from the nanocluster. By turning off the source, the final length of the nanowire can be adjusted.

The nanowires of a given material exhibit thermal, electrical, mechanical and optical properties different from the bulk material. For example, the conductivity of a nanowire will be much less than that of the corresponding bulk material.

Uses

- Data storage/transfer - transfer data up to 1,000 times faster, and store data for as long as 100,000 years without degradation.
- Batteries/generators - tiny, efficient solar panels, turning light into energy, able to hold 10 times the charge of existing batteries.

- Transistors.
- LED's.
- Optoelectronic devices.
- Biochemical sensors.
- Thermoelectric devices.

(e) Nanorods

Nanorods are solid nanostructures morphologically similar to nanowires but with aspect ratios (length divided by width) of approximately 3-to-5. They are formed from a variety of materials including metals, semiconducting oxides, diamonds (aggregated diamond nanorods produced from fullerenes are the hardest material so far discovered), and organic materials. Nanorods are produced by a number of techniques, including: a vapor-liquid-solid approach; mechanical alloying; direct chemical synthesis; plasma arc discharge; laser ablation; and catalytic decomposition.

Direct chemical synthesis method is one of the simple and most commonly used method for the synthesis of nanorods, with a combination of ligands acting as shape control agents. The ligands bond to different facets of the nanorod with different strengths. This allows different faces of the nanorod to grow at different rates, producing an elongated object.

Uses

- In display technologies, because the reflectivity of the rods can be changed by changing their orientation with an applied electric field.
- In microelectromechanical systems (MEMS).
- Nanorods, along with other noble metal nanoparticles, also function as theragnostic agents.
- As cancer therapeutics. Nanorods absorb in the near IR, and generate heat when excited with IR light. Nanorods can be conjugated with tumor targeting motifs and ingested. When a patient is exposed to IR light (which passes through body tissue), nanorods, selectively taken up by tumor cells are locally heated, destroying only the cancerous tissue while leaving healthy cells intact.
- Nanorods based on semiconducting materials as energy harvesting and light emitting devices.

(f) Dendrimers

Dendrimers are large and complex molecules with very well-defined chemical structures. The dendrimers are tree-like repetitively branched polymer molecules (Greek dendra = tree). They are also termed **cascade molecules**. They have particle size in the nanometer range. They are branching molecules with the branching beginning at the core. Depending on the core, the dendrimer can start with 3 to 8 (or more) branches, with 3 and 4 being the most common number. Figure 9.9 schematically represents a dendrimer.

Dendrimers consist of a core molecule and alternating layers of two monomers. Each pair of monomer layers completes a shell and a generation. The core generally consists of an amine core, although sugars and other molecules can be used. All core molecules have multiple reaction sites that are identical. Even the simplest core possible, ammonia (NH_3) has three amine reaction sites.

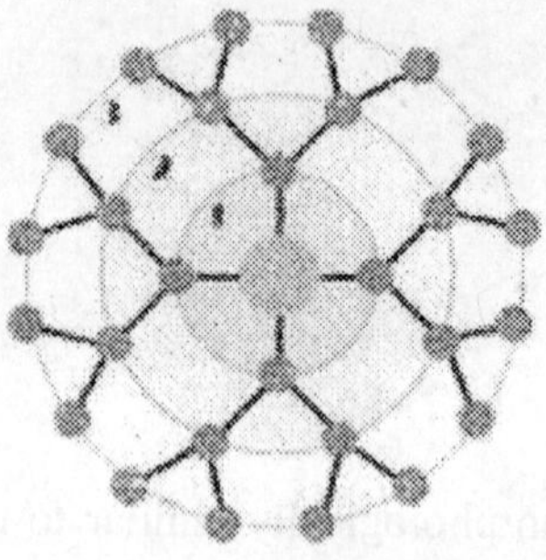

Fig. 9.9. Dendrimer

The core is mixed with an excess of the first monomer molecule which reacts with all of the core's reaction sites, giving rise to the first branches. This monomer molecule has two distinct reactive groups, one at each end. After one kind of end reacts, the other end will provide reaction sites for the next layer of the shell. An excess of the second monomer, again a molecule which has two distinct reactive groups, one at each end, is reacted with this first layer to give the second layer and complete the first shell and the first generation. Each unreacted outer end of the second monomer provides a reaction site which can react with multiple molecules. This provides the branching and the reactions sites for the next shell.

Uses

Applications of dendrimers typically involve conjugating other chemical species to the dendrimer surface that can function as detecting agents (such as a dye molecule), affinity ligands, targeting components, radioligands, imaging agents, or pharmaceutically active compounds. Such molecules could include tumor-targeting agents (including but not restricted to monoclonal antibodies), imaging contrast agents to pinpoint tumors, drug molecules for delivery to a tumor, and reporter molecules that might detect if an anticancer drug is working.

In general, the dendrimers can be used in variety of applications as:

- Medicinal and diagnosis applications,
- Gene therapy and chemical sensors,
- Drug delivery system,
- Adhesive and coatings,
- Light harvesting material,
- Catalyst, electronic applications,
- Separating agents, and many more.

(g) Nanocomposites

Nanocomposites are a class of materials in which one or more phases with nanoscale dimensions (0-D, 1-D, and 2-D) are embedded in a metal, ceramic, or polymer matrix. According to their matrix materials, nanocomposites can be classified as ceramic matrix nanocomposites (CMNC), metal matrix nanocomposites (MMNCs), and polymer matrix nanocomposites (PMNCs). The reinforcing material can be made up of nanoparticles, nanosheets or nanofibres.

The mechanical, electrical, thermal, optical, electrochemical, catalytic properties of the nanocomposites differ markedly from that of the component materials. For example, adding carbon nanotubes improves the electrical and thermal conductivity of the composite. Other kinds of nanoparticulates may result in enhanced optical properties, dielectric properties, heat resistance or mechanical properties such as stiffness, strength and resistance to wear and damage.

The nanocomposites differ from conventional composite materials due to the exceptionally high surface to volume ratio of the reinforcing phase and/or its exceptionally high aspect ratio. The area of the interface between the matrix and reinforcement phase(s) is typically an order of magnitude greater than for conventional composite materials. The matrix material properties are significantly affected in the vicinity of the reinforcement. The large amount of reinforcement surface area suffices a relatively small amount of nanoscale reinforcement for an observable effect on the macroscale properties of the composite. The percentage by weight (called *mass fraction*) of the nanoparticulates introduced can remain very low (on the order of 0.5% to 5%).

Uses

- **In producing** batteries **with greater power output.** In lithium ion batteries the anodes made of the silicon-carbon nanocomposite make closer contact with the lithium electrolyte, which allows faster charging or discharging of power.
- In **speeding up the healing process for** broken bones. The growth of replacement bone is speeded up when a nanotube-polymer nanocomposite is placed as a kind of scaffold which guides growth of replacement bone.
- **In producing structural components with a high strength-to-weight ratio.** For example, an epoxy containing carbon nanotube can be used to produce nanotube-polymer composite windmill blades. This results in a strong but lightweight blade, which makes longer windmill blades practical. These longer blades increase the amount of electricity generated by each windmill.
- In **making lightweight sensors with nanocomposites.**
- **In making flexible batteries.** A nanocomposite of cellulous materials and nanotubes could be used to make a conductive paper. When this conductive paper is soaked in an electrolyte, a flexible battery is formed.
- In **making** tumors **easier to see and remove.** Incorporation of magnetic nanoparticles and fluorescent nanoparticles in a nanocomposite makes it both magnetic and fluorescent. The magnetic property of the nanocomposite particle makes the tumor more visible during an MRI procedure done prior to surgery. The fluorescent property of the nanocomposite particle could help the surgeon to better see the tumor while operating.
- As light weight materials in vehicles, for fuel economy.
- As artificial joints, economically beneficial.
- In abrasion and wear resistant applications.

REVIEW QUESTIONS

1. Define a nanomaterial.
2. Explain the classification of nanomaterials with examples.
3. Why do nanomaterials exhibit properties different from bulk materials?
4. Give an account on the following size dependent properties of nanomaterials:
   i. Thermal properties
   ii. Electrical properties
   iii. Optical properties
   iv. Magnetic properties
   v. Mechanical properties.
5. Differentiate between bottom-up and top-down approach in the synthesis of nanomaterials.
6. Explain the following methods of nanomaterial synthesis:
   i. Sol-gel process
   ii. Precipitation process
   iii. Gas condensation process
   iv. Chemical vapor condensation method
   v. Hydrothermal process
   vi. Thermolysis process.
7. Write notes on the following:
   i. Fullerenes
   ii. Carbon nanotubes
   iii. Nanowires
   iv. Nanorods
   v. Dendrimers
   vi. Nanocomposites.
8. Enumerate the uses of the following:
   i. Fullerenes
   ii. Carbon nanotubes
   iii. Nanowires
   iv. Nanorods
   v. Dendrimers
   vi. Nanocomposites.

Laboratory Manual
in
Engineering Chemistry

Part A

1. POTENTIOMETRIC TITRATION

Aim

To estimate the weight of Mohr's salt dissolved in a liter of the given solution by potentiometric titration with standard potassium dichromate solution.

Principle

A potentiometric titration may be defined as a titration in which the end point is detected by measuring the change in potential of a suitable electrode (which responds to the change in concentration) during the titration. The electrode which responds to the change in concentration of the ion in the solution is called the indicator electrode. The indicator electrode is combined with a reference electrode (whose potential does not change during the titration) to form a cell and e.m.f. of the so formed cell is measured during the titration. The e.m.f. of the cell changes gradually till the end point and changes rapidly at very close to the end point and again the change is gradual after the end point. When e.m.f., E is plotted as ordinate and the volume of titrant added as abscissa, the point of inflection of the curve corresponds to the equivalence point or end point of the titration.

A better method is by plotting the first derivative curve, in which ($\Delta E/\Delta V$) is plotted against V, and the end point is obtained at the maximum of the curve.

The titration of Mohr's salt solution with $K_2Cr_2O_7$ in the presence of H_2SO_4 is a redox titration.

$$3Fe^{2+} + Cr^{6+} \rightarrow 3Fe^{3+} + Cr^{3+}$$

The presence of oxidized and reduced forms of the same substance in a solution gives rise to the formation of an oxidation-reduction electrode, developing an electrode potential, which can be picked up by dipping a Pt wire. Thus when the titration is commenced, both Fe^{2+} and Fe^{3+} ions are present in the solution, developing an electrode potential which is picked up by a Pt wire. Thus (Pt/Fe^{3+}, Fe^{2+}) is the indicator electrode, whose electrode potential is given by,

$$E = E^\circ + \frac{2.303RT}{F} \log \frac{[Fe^{3+}]}{[Fe^{2+}]}$$

It is combined with a reference electrode (calomel if potentiometer is used and glass electrode if pH meter is used) and the e.m.f. of the cell is measured.

During the titration $[Fe^{3+}]$ goes on increasing and $[Fe^{2+}]$ goes on decreasing as $K_2Cr_2O_7$ solution is added continuously, gradually changing the potential. Near the end point the ratio $[Fe^{3+}]/[Fe^{2+}]$ increases rapidly as $[Fe^{2+}]$ becomes very small, and tending to zero at the end point. Thus the e.m.f. increases rapidly at very close to the end point. When Fe^{2+} is completely converted into Fe^{3+} by $K_2Cr_2O_7$ the electrode, Pt/Fe^{3+},Fe^{2+} ceases to exist. But the presence of slight excess of $K_2Cr_2O_7$ brings in existence of Pt/Cr^{6+},Cr^{3+} electrode. Thus after the end point, it is, the electrode potential of the Pt/Cr^{6+},Cr^{3+} electrode, which is going to change. Because of both the factors, i.e., increase in the value of $[Fe^{3+}]/[Fe^{2+}]$ and change of electrode from Pt/Fe^{3+},Fe^{2+} to Pt/Cr^{6+},Cr^{3+}, there is a large change in potential at the end point.

PROCEDURE

Pipette out 25ml of the given Mohr's salt solution into a 250ml beaker. Add a test tube of dil H_2SO_4. Dip a platinum electrode and a reference electrode (glass electrode for pH meter or calomel electrode for potentiometer). Add sufficient distilled water so that the electrodes are properly dipped in the solution. Connect the electrodes to a potentiometer (or pH meter). Go on adding 1ml of $K_2Cr_2O_7$ solution from a burette, stir well with the help of a glass rod and note down the e.m.f. of the cell from the meter for each addition. Continue the titration till a sharp increase in potential is obtained and 5–6 readings there after.

Clean the beaker and the electrodes and repeat the titration once again by adding $K_2Cr_2O_7$ solution directly into the beaker till 0.5ml less than the lower range of end point and then adding 0.1 ml at a time. Continue till a sharp change in potential at the end point for 0.1 ml addition of $K_2Cr_2O_7$ solution, and 5–6 more readings there after. Plot $\Delta E/\Delta V$ (Y-axis) vs V (X-axis) and note down the end point.

RESULT

Weight of Mohr's salt in a liter of the solution = ...

OBSERVATIONS ARID CALCULATION

Concentration of the given $K_2Cr_2O_7$ solution, N_1 = ...

TITRATION OF MOHR'S SALT SOLUTION WITH $K_2Cr_2O_7$ SOLUTION

Vol of $K_2Cr_2O_7$ added in ml	*EMF, E(mV)*	*ΔV*	*ΔE*	*ΔE/ΔV*

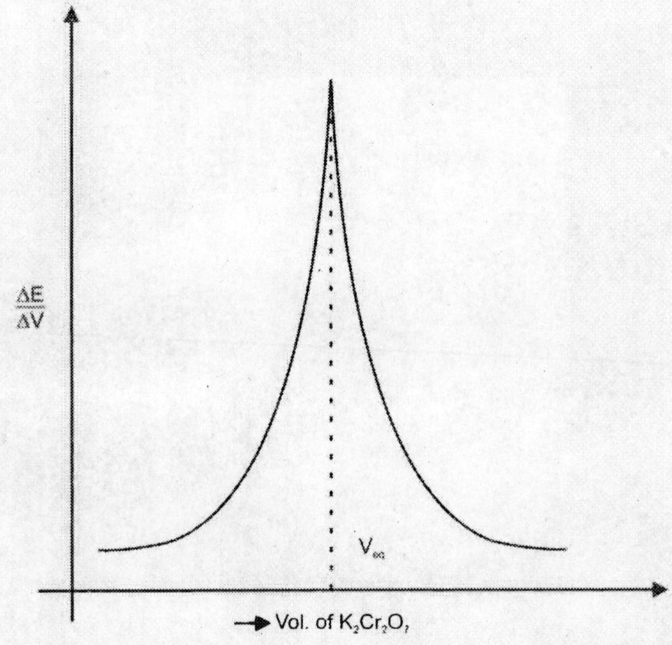

Fig. 1.

From the graph, the volume of $K_2Cr_2O_7$ solution consumed, V_1 = ... ml

$$V_1 \times N_1 = V_2 \times N_2$$

Concentration of Mohr's salt solution, $N_2 = \dfrac{V_1 \times N_1}{50}$

Weight of Mohr's salt in a liter of the solution= N_2 × eq. wt. of Mohr's salt

= N_2 × 392

= ....g

Signature of the staff member

2. COLORIMETRIC ESTIMATION OF COPPER

Aim

To estimate the amount of copper present in the given solution by colorimetric method.

Principle

Colorimetric analysis depends upon the measurement of quantity of light absorbed by a colored solution. Quantitative analysis by colorimetry is based on Beer-Lambert Law, which can be expressed by the relation.

$$\log (I_o/I_t) = \varepsilon cl$$

Where

I_o = intensity of the incident light
I_t = intensity Of the transmitted light
c = concentration of the colored constituent in the solution
l = thickness of the colored medium and
ε = molar absorption coefficient

The term, $\log (I_o/I_t)$ is called the absorbance or optical density of the light-absorbing medium. Since ε is a constant and if thickness l is kept constant, absorbance changes linearly with concentration within a specific concentration range.

The ratio (I_t/I_o) is the transmittance, T of the solution.

Therefore, absorbance, $\log (I_o/I_t) = \log(1/T) = -\log T$

So, the same result can be obtained by plotting log T against concentration also.

Cupric ions (Cu^{2+}) react with ammonia to give a deep blue color due to the formation of cuprammonium complex ion.

$$Cu^{2+} + 4NH_3 \rightarrow \underset{\text{Deep blue}}{[Cu(NH_3)_4]^{2+}}$$

A series of standard solutions containing cupric ions are treated with ammonia and the absorbance of each of these solutions is measured at 620nm (λ_{max}, at which the solution shows maximum absorbance). A calibration curve is obtained by plotting concentration against absorbance or log T. A known volume of the sample solution is also treated with ammonia in the same way as the standards and its absorbance is measured. Its concentration is determined from the calibration curve.

Procedure

Pipette out 5,10,15,20 and 25ml of the standard copper sulfate solution (containing l mg of copper per ml) into 50ml standard flasks. Add 10ml of 1 : 1 ammonia solution into each of them and make up to the mark with distilled water and shake well.

Insert a filter (620 nm) (if spectrophotometer is used select the wave length of 620 nm) in the colorimeter and adjust the initial readings to 100% transmittance (or zero absorbance) with distilled water as blank in the sample tube. Fill the sample tube with the standard solution of lowest concentration and measure the absorbance and transmittance. Repeat for other standard solutions also. Plot absorbance or log T (Y-axis) versus concentration (X-axis).

Develop color for the given unknown solution by adding 10 ml of 1: 1 ammonia, and make up to the mark with distilled water. Measure the absorbance and transmittance using the colorimeter at the same wavelength. From the graph, determine the concentration of the unknown solution.

Result

Weight of copper (II) present in the given solution = … mg

Observations and Calculation

Weight of Cu(II) per ml of the solution = …

Vol. of Cu(II) solution pipetted out in ml	*Concentration of Cu(II), mg*	*Absorbance*	*Transmittance (A)*	*logT (T)*
5				
10				
15				
20				
Unknown				

From the graph, the unknown solution contains ………….mg of Cu(II)

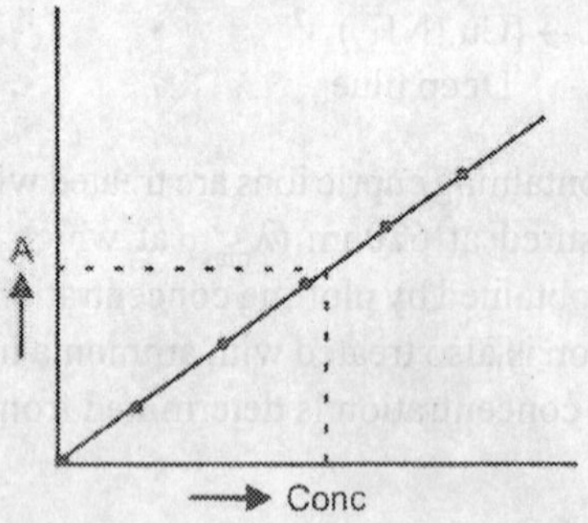

Fig. 2.

Signature of the staff member

3. CONDUCTOMETRIC TITRATION – ACID BASE TITRATION

Aim

To determine the concentration of a weak acid and a strong acid in a mixture by conductometric titration with a strong base.

Principle

In conductometric titrations, the conductance of the solution is measured during titration, before and after the equivalence point, and the variation of conductance of the solution during the titration is utilized to locate the end point.

In a mixture of strong acid and a weak acid, the strong acid undergoes complete dissociation as it is a strong electrolyte and the weak acid remains almost undissociated. When a mixture of strong acid and a weak acid is titrated against a strong base, the strong acid is consumed first by the base. After the neutralization of the strong acid is completed, then only the weak acid is neutralized. When the strong acid is neutralized by the base, H^+ ions of the acid are consumed and are replaced with metal ions in the solution. For example for HCl + NaOH,

$$H^+ + Cl^- + Na^+ + OH^- \rightarrow Na^+ + Cl^- + H_2O$$

Since highly conducting H^+ ions are replaced by weakly conducting Na^+ ions during the titration, the conductance of the solution sharply decreases. When the weak acid is neutralized, the poorly conducting acid is converted into a salt. The salt undergoes complete dissociation, and therefore, the conductance increases gradually. After the neutralization of the weak acid is also completed, further addition of NaOH increases the conductivity sharply in proportional with the amount of NaOH added, as more and more Na^+ and OH^- ions are added into the solution. Therefore if the conductance of the solution is plotted against volume of the base added we will get three straight lines. The points of intersection of the lines give the equivalence points corresponding to the strong acid and the total acid respectively, and the equivalence point for the weak acid can be deduced.

The solution taken in the burette should be at least ten times more concentrated than the analyte solution to avoid the effect of dilution on the variation of conductance.

Procedure

Pipette out 50 ml of the given acid mixture solution into a 100 ml beaker, inserted with a glass rod. Clean the conductance cell with distilled water, dry it with filter paper strips and dip in the acid solution. Connect the conductivity cell to the conductometer and select a suitable range for conductance, so that, maximum number of digits is displayed on the display of the conductometer. Fill a burette with the given strong base, NaOH. Add 0.5 ml of the base into the beaker at a time, stir well and note the conductance of the solution after each addition. The titration should be continued till the decreasing trend of conductance changes to a gradually increasing trend, and then to a relatively sharply increasing trend; and at least 5-6 readings should be taken after that. Plot volume of the base added along the x-axis and conductance along the y-axis. The points of intersection of the curves give the equivalent points, the first point of intersection corresponding to the strong acid and the second corresponding to the total acid in the solution, respectively.

Result

(i) Equivalence volume for the Strong acid : V_1
(ii) Concentration of the strong acid in the given solution :
(iii) Equivalence point for weak acid : V_3
(iv) Concentration of the weak acid in the given solution :

Observations and Calculations:

Concentration of the base $= N_1 =$
Volume of the acid mixture pipetted out $= 50\ cm^3$

Vol. of base added (cm^3)	*Conductance*	*Vol. of base added (cm^3)*	*Conductance*

Equivalence volume for the strong acid from the graph $= V_1 = \ldots\ldots\ldots\ldots cm^3$

$\therefore$ Concentration of the strong acid $\dfrac{V_1 \times N_1}{50}$

$\ldots\ldots\ldots\ldots$ N

Equivalence volume for the total acid from the graph

$\therefore$ Equivalence volume for the weak acid, $V_3 = (V_2 - V_1)$

$\therefore$ Concentration of the weak acid $\dfrac{(V_2 - V_1) \times N_1}{50}$

$= \ldots\ldots\ldots\ldots$ N

Signature of the staff member

4. FLAMEPHOTOMETRY

Aim

To estimate the sodium and potassium content in the given sample of water by flamephotometry.

Principle

Emission of characteristic radiation by an element and the correlation of the emission intensity with the concentration of the element form the basis of **flame photometry.**

When a solution containing the sample element or ion is aspirated into the flame, a series of changes take place at the flame. First, the solvent gets evaporated leaving behind the salt in the flame. The salt then gets evaporated into vapors of the salt, which further undergo dissociation into the constituent atoms. Some of the metal atoms formed in the flame may absorb heat energy from the flame and get electronically excited to their higher energy level. Being unstable at the excited level, the atoms fall back to their ground state by emitting the energy difference between the excited and ground state, in the form light radiation. The intensity of the light radiation emitted is proportional to the number of atoms in the excited state, which in turn is proportional to the number of atoms in the flame or the concentration of the solution fed into the flame. Thus concentration of the solution is related to the intensity of the emitted radiation. By measuring the intensity of the emitted radiation by a suitable detector device, the concentration can be determined. As different metals emit their characteristic radiations at different wave lengths, they do not interfere with each other, even when they are present together. The different radiations emitted by the different metal atoms can be resolved by using a dispersive unit, detected and measured for analysis.

The sequence of changes taking place at the flame, when metal salt solution is aspirated may be summarized as follows:

$$\underset{\text{Solution}}{M^+X^-} \xrightarrow{\text{Evaporation}} \underset{\text{Mist}}{M^+X^-} \xrightarrow{\text{vaporization}} \underset{\text{Solid}}{MX} \xrightarrow{\text{dissociation}} \underset{\text{Gas}}{MX} \rightarrow M(gas) + X(gas)$$

$$M(gas) \xrightarrow{\text{absorption of heat}} M^*(gas) \xrightarrow{\text{Flame emission, } h\upsilon} M\,(gas)$$

The intensity of the emitted radiation, measured as detector response is related to the concentration by an expression similar to Beer's relation,

$$E = k\alpha c,$$

where E is the detector response, k is a constant, α is the efficiency of atomic excitation and c is the concentration of the solution.

The quantitative analysis by flame photometry can be carried out by calibration curve method. For this purpose, a series of standard solutions of the analyte metal are taken, aspirated into the flame and the emission of each solution is measured in a flame photometer. Then the calibration curve is obtained by plotting the emission intensity against the concentration of the standard solutions. The sample solution is properly diluted so that its emission is in the range of the standards, and its emission is measured. From the calibration curve, the concentration of the sample solution can be determined.

Procedure

Pipette out 5, 10, 15, and 20ml of the standard NaCl solution containing 0.1 mg of Na per ml of the solution into 4 different 100ml standard flasks, make the solution up to the mark with distilled water and shake well for uniform concentration. Take the first standard solution in a small beaker and aspirate into the flame of the flamephotometer. The flame emission intensity of the standard is measured by the flamephotometer using sodium filter (598 nm) in the photometer. The same is repeated for all the standard solutions with distilled water being sprayed to the flame in between the trials. Plot the calibration curve by plotting emission response along the Y-axis and concentration of the solution along the X-axis. Similarly get a calibration curve for potassium by using potassium sulphate solution containing 0.1 mg of K per ml of the solution. Flame emission intensity to be measured using the potassium filter.

Then aspirate the given sample of water into the flame and note down the value of emission response using both sodium and potassium filter. From the calibration curves compute the concentrations of Na and K in the water sample.

Result

Amount of sodium present in the given sample of water = ...

Observations and Calculation

(i) Weight of sodium per ml of the solution = ...

Vol. of NaCl solution pipetted out in ml	*Concentration of Na, mg*	*Emission response*
5		
10		
15		
20		
Unknown		

From the graph, the unknown solution contains .... mg of sodium.

Signature of the staff member

(ii) Weight of potassium per ml of the solution = ...

Vol. of K_2SO_4 solution pipetted out in ml	*Concentration of K, mg*	*Emission response*
5		
10		
15		
20		
Unknown		

From the graph, the unknown solution contains .... mg of potassium.

Signature of the staff member

5. DETERMINATION OF pKa VALUE OF A WEAK ACID

Aim

To determine the pKa value of a weak acid by pH titration method.

Principle

Weak acid is a weak electrolyte and undergoes only partial dissociation in the solution. For example, acetic acid when dissolved in water

$$CH_3COOH \rightleftharpoons CH_3COO^- + H^+$$

The dissociation constant, $Ka = \dfrac{[CH_3COO^-][H^+]}{[CH_3COOH]}$

$$pKa = -\log Ka$$

Higher the value of pKa, lower is the Ka value of the acid and therefore weaker is the acid.

pKa value of a weak acid can be determined by pH titration method. In this method a known volume of the weak acid is titrated with a strong base, and pH of the solution is measured after each small addition of the base. During the titration, the pH of the solution increases gradually till very near to equivalence point and there is a sharp increase in pH at the equivalence point. After the equivalence point the increase in pH is again very gradual. Therefore the equivalence point is indicated by the sharp increase in pH (Fig. 3). The equivalence point is determined by plotting volume of the base against ΔpH/ΔV, as indicated by the volume corresponding to the maximum of the curve (Fig. 4)

Once the titration is started, the solution contains weak acid and also it's salt with strong base. Thus the mixture is a buffer mixture. The pH of such a mixture is give by Hendersen's equation;

$$pH = pKa + \log\frac{[salt]}{[acid]}$$

At half equivalence point 50% of the acid is converted into salt, and therefore, [acid] = [salt].

Then pH = pKa.

Therefore pH at half equivalence point gives the pKa value of the weak acid.

Procedure

Pipette out 25ml of the given weak acid into a beaker. Dip the glass electrode – calomel electrode assembly into the acid and connect the cell to a pH metre. Measure the pH of the solution. Add 1 ml of NaOH at a time into the beaker, stir the solution and measure the pH of the solution. Continue the process till a large increase in pH is observed at the end point for the addition of 1 ml of NaOH solution and take four more readings after that. Use the readings to plot a graph of volume of base against pH.

Now clean the beaker well and again pipette out 25ml of the weak acid. Add NaOH solution, 0.5 ml less than the lower range of the end point directly and then 0.1 ml of NaOH at a time. After each addition, stir the solution and measure the pH of the solution. Continue the titration till a sharp increase in pH is observed for the addition 0.1 ml of NaOH at the end point. Continue the titration further to note down at least four more readings. Plot volume of NaOH against $\Delta pH/\Delta V$ and determine the equivalence point. From the graph of volume of NaOH against pH, determine the value of pKa as pH at half equivalence point.

Result:

(i) Equivalence point of the titration = ... ml
(ii) pKa value of the given weak acid = ...

Observations and Calculation

Volume of NaOH added (ml)	*pH*	*Volume of NaOH added (ml)*	*pH*	ΔV	ΔpH	$\frac{\Delta pH}{\Delta V}$
1						
2						
3						
4						
5						
6						
7						
8						
9						
10						
11						

From the graph, equivalence point of the titration = ...ml

pH at half equivalence point = ...

∴ pKa of the given weak acid = ...

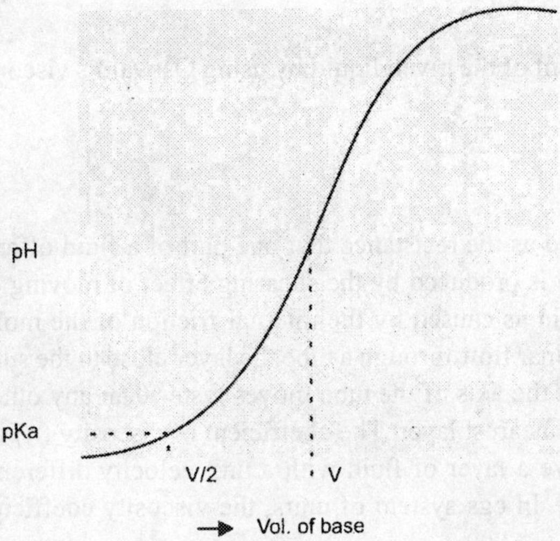

Fig. 3.

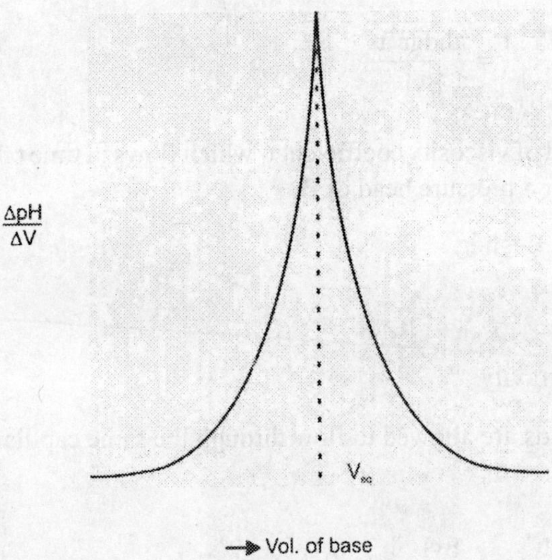

Fig. 4.

Signature of the staff member

6. DETERMINATION OF VISCOSITY COEFFICIENT OF A LIQUID

Aim

To determine the viscosity coefficient of the given liquid by using Ostwald's viscometer.

Principle

Viscosity of a liquid may be defined as the resistance that one part of a fluid offers to the flow of another part of the liquid. Viscosity is produced by the shearing effect of moving one layer of the fluid past another. It may be thought as caused by the internal friction of the molecules between themselves. When a liquid is in laminar flow through a tube the layer close to the surface of the tube is almost stationary and the layer at the axis of the tube moves faster than any other layer. A slow moving layer exerts a friction on its nearest layer. The coefficient of viscosity (η) is defined as the force per unit area required to move a layer of fluid with a unit velocity difference past another parallel layer at unit distance away. In cgs system of units, the viscosity coefficient of a fluid is expressed in poises.

The viscosity coefficient of a liquid is given by Poiseuille's equation.

$$\eta = \frac{\pi P r^4 t}{8vl} = \frac{\pi h d g r^4 t}{8vl}$$

where v is the volume of the liquid of viscosity coefficient η which flows in time t through a capillary tube of radius r and length l under a pressure head of P.

$$P = hdg$$

where h = height
d = density
g = acceleration due to gravity.

If equal volumes of two liquids are allowed to flow through the same capillary under identical conditions,

$$\frac{\eta_1}{\eta_2} = \frac{\pi h d_1 g r^4 t_1}{8vl} \times \frac{8vl}{\pi h d_2 g r^4 t_2} = \frac{d_1 t_1}{d_2 t_2}$$

The flow times for the liquid and water are determined in a Ostwald's viscometer (Fig. 5). Knowing the densities of the liquid and water and also knowing the viscosity coefficient of water, viscosity coefficient of liquid can be calculated.

Since viscosity is dependant on temperature, the experiment has to be carried out at constant temperature. Therefore viscometer should be dipped in a jar containing water in such a way that the

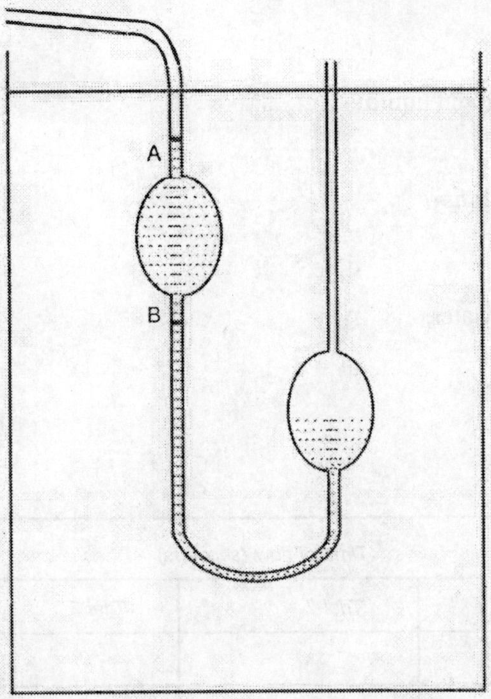

Fig. 5.

mark above the upper bulb is always below water, so that, the liquid level inside the viscometer is inside water. Small variation in atmospheric temperature will not be felt by the liquid when the viscometer is inside the water.

Procedure

Take a clean and dry viscometer and fix it vertically to a stand in water taken in a jar, in such a way that the mark above the upper bulb is below the water level. Pipette out a definite known volume (10 ml or 20 ml depending upon the size of the upper bulb) of the liquid into the lower bulb. With the help of rubber tubing fitted to the limb containing the upper bulb, suck the liquid into the upper bulb and above the mark A of the viscometer. Allow it to flow freely through the capillary. When the level of the liquid just crosses the mark A, start a stop clock. Stop the clock when the liquid crosses the lower mark B. Note the flow time in seconds. Repeat for two more times.

Remove the liquid from the viscometer, rinse with acetone and dry it. Now pipette out the same known volume of distilled water as before into the lower bulb of the viscometer and find out the flow time as before.

Result

Viscosity coefficient of the given liquid = ...

Observations and Calculations

Density of water, $d_1 =$

Viscosity coefficient of water, $\eta_1 =$

Density of the liquid, $d_2 =$

To determine the flow time

	Time of flow (seconds)			
	Trial 1	*Trial 2*	*Trial 3*	*Average*
Liquid				
Water				

$$\frac{\eta_1}{\eta_2} = \frac{d_1 t_1}{d_2 t_2}$$

$\therefore$ Viscosity coefficient the liquid, $\eta_2 = \frac{d_1 t_1}{d_2 t_2} \times \eta_1$

$=$

$=$ ....

Signature of the staff member

Part – B

1. ESTIMATION OF HARDNESS OF WATER

Aim

To estimate the total hardness of the given sample of water.

Principle

Hardness of water is due to the presence of dissolved salts of calcium, magnesium and other metal ions. Hard water does not give lather easily with soap, as the hardness causing ions react with soap to form insoluble soap. Temporary hardness is due to the bicarbonates of calcium and magnesium, which can be removed by boiling the water. On boiling, the bicarbonates are converted into insoluble carbonates of calcium and magnesium. Permanent hardness is due to the presence of other soluble salts of calcium, magnesium and other metals, which can be removed only by chemical treatment. The sum of temporary hardness and permanent hardness is known as total hardness. The hardness is expressed in ppm equivalent of calcium carbonate or mg/l.

Hardness of water is determined by titrating a known volume of water sample with EDTA, using Eriochrome Black T (EBT) as indicator, in the presence of ammonium hydroxide – ammonium chloride buffer (pH 10). Ethylenediamine tetra acetic acid (EDTA) or its disodium salt is a strong complexing agent, reacts quantitatively with the metal ions to form a soluble, stable, 1: 1 complex.

$$(NaOOCCH_2)_2NCH_2CH_2N(CH_2COOH)_2 + M^{2+} \rightarrow (NaOOCCH_2)NCH_2CH_2N(CH_2COO)_2M + 2H^+ \quad (M = Ca^{2+}, Mg^{2+})$$

When a few drops of the indicator EBT are added, it forms a wine red colored complex with the metal ions.

$$M^{2+} + EBT \rightarrow \underset{\text{(Wine red)}}{M\text{–}EBT \text{ complex}}$$

When the solution is titrated, EDTA reacts, preferentially with free metal ions present in the solution. Near the end point, when the free metal ions are exhausted in the solution, further addition of EDTA dissociates the M–EBT complex, consumes the metal ions and releases free indicator, which is blue in color. Therefore the color change is from wine red to blue.

The indicator, EBT shows the color change at pH range 9–11. Therefore, the pH of the solution is maintained around 10 using $NH_4OH + NH_4Cl$ buffer.

EDTA being insoluble in water, it is generally used as a disodium salt which readily dissolves in water.

A standard solution of EDTA is prepared by dissolving a known weight of EDTA in distilled water and making up to 250 ml.

Molecular weight of disodium salt of EDT A = 372.24

$$\text{Molarity of EDT A solution} = \frac{\text{wt. taken}}{\text{Mol. wt}} \times 4$$

Procedure

Preparation of standard solution of EDTA: Weigh accurately about 2.5 g of EDTA and transfer into a 250 ml standard flask. Dissolve completely with distilled water (add few drops of NaOH solution if the crystals do not dissolve). Make up the solution to the mark and shake well for uniform concentration.

Estimation of Hardness of Water: Pipette out 25 ml of hard water into a clean conical flask. Add 2ml of buffer solution, 1 t.t. of distilled water and 2–3 drops of Eriochrome Black T indicator. Titrate against EDTA solution taken in the burette until the color of the solution changes from wine red to blue . Repeat the titration for concordant values.

Result

Total hardness of the given sample of water =........ ppm (parts of $CaCO_3$ per million parts of water).

Observations and Calculation

Preparation of standard EDTA solution:

Wt of w. bottle + EDTA,	W_1
Wt of w. bottle only,	W_2
Wt of EDTA = $(W_1 - W_2)$,	W

Molarity of EDT A solution, $M_1 = \frac{W \times 4}{372.24} =$

= ....

Estimation of total hardness of water

In burette : EDTA solution
In conical flask : 25ml water sample + 2ml buffer solution + 1.t.t distilled water
Indicator : Eriochrome Black T, 2–3 drops
Color change : Wine red to blue

Burette level	*I*	*II*	*III*
Final			
Initial			
Vol of EDTA, ml			

Agreeing value, V_1 = ... ml

25 ml of Hard water consumes V_1 ml of M_1 M EDTA solution.

Since $V_1M_1 = V_2M_2$ and $V_2 = 25$ ml,

Molarity of hard water, $M_2 = \frac{V_1M_1}{25}$

= ....

∴ Wt of calcium carbonate present in a liter of the solution

$W_3 = M_2 \times \text{Mol wt of } CaCO_3$

$= M_2 \times 100$,

= .... = ... g

Hardness of water $= \frac{W_3 \times 10^6}{1000}$

= .... ppm

Signature of the staff member

2. ESTIMATION OF CALCIUM OXIDE IN CEMENT

Aim

To estimate the amount of calcium oxide present in the given sample of cement solution.

Principle

Cement contains compounds of calcium, aluminium, magnesium, iron and insoluble silica. When dissolved in acid, silica remains undissolved. On treating with ammonia, aluminium and iron can be precipitated as their hydroxides and separated. The provided solution of cement, therefore, contains calcium and magnesium ions.

To estimate the calcium content in the given solution, a known volume of it is titrated with standard EDTA solution. In the presence of magnesium, calcium can be titrated with EDTA by using Patton and Reeder's indicator at pH values between 12 and 14. The indicator combines with calcium ions forming a red colored complex.

$$Ca^{2+} + In \rightarrow Ca - In \text{ complex}$$

(Wine red)

Near the end point, when free calcium ions are exhausted in the solution, further addition of EDTA dissociates Ca-In complex, consumes the calcium ions and releases free indicator, which is blue in color. Therefore, the color change is wine red to blue.

The interference of magnesium ions is avoided by precipitating them as hydroxide. This is done by adding 8N NaOH and pH of the solution is maintained at a pH of 12.5 by adding 5 ml of diethylamine. If the indicator is added before the NaOH solution, satisfactory end point is not obtained because magnesium salts form a lake with the indicator as the pH increases and the magnesium – indicator lake is co-precipitated with the magnesium hydroxide. A sharper end point may be obtained by adding 2–3 drops of 1% aqueous poly (vinyl alcohol) to the sample solution before adding NaOH solution. The poly (vinyl alcohol) reduces the adsorption of the dye on the surface of the precipitate.

The standard EDTA solution is prepared by dissolving a known weight of disodium salt of EDTA in a known volume of the solution.

Molecular weight of disodium salt of EDTA = 372.24

Procedure

Preparation of standard solution of EDTA: Weigh accurately about 2.5g of EDTA and transfer into a 250 ml standard flask. Dissolve completely with distilled water (add a few drops of NaOH solution if the crystals do not dissolve). Make up the solution to the mark and shake well for uniform concentration.

Estimation of calcium oxide in cement solution: Pipette out 25ml of the given cement solution into a clean conical flask. To the flask add 2–3 drops of 1% aqueous poly (vinyl alcohol), and 5 ml of diethylamine and 5 ml of 8M NaOH solution. Allow the solution to stand for 3–5 minutes with occasional stirring. Add a test tube of distilled water, a pinch of Patton and Reeder's indicator and titrate against the EDTA solution taken in the burette till the color changes from wine red to blue. Repeat for agreeing values.

Result

Weight of calcium oxide on a liter of the given sample of cement solution = … g.

Observations and Calculations

Preparation of standard EDTA solution

Weight of w. bottle + EDTA, W_1 =

Weight of w. bottle alone, W_2 =

Weight of EDTA, $(W_1 - W_2)$, W =

Molarity of EDT A solution, $M_1 = \frac{W \times 4}{372.24} =$

= …………..

Estimation of CaO

In burette : EDTA solution
In conical flask : 25ml cement solution + 2–3 drops of aq. poly (vinyl alcohol) + 5ml of diethyl amine + 5ml 8N NaOH + 1 t.t. distilled water
Indicator : Patton and Reeder's indicator, a pinch
Colour Change : Wine red to Blue

Burette Level	*I*	*II*	*III*
Final			
Initial			
Volume of EDTA, ml			

Agreeing value, V_1 =…………ml

25ml of cement solution consumes V_1 ml of M_1 M EDTA solution.

$$V_1M_1 = V_2M_2$$

Molarity of cement solution, $M_2 = \frac{V_1M_1}{25} =$ ....

= ....

Wt of calcium oxide in a litre of the cement solution, $W = M_2 \times$ Mol wt of CaO

$= M_2 \times 56.08$

= ...

= ...

Signature of the staff member

3. ESTIMATION OF COPPER IN BRASS

Aim

To estimate the percentage of copper in the given sample of brass, being supplied with standard sodium thiosulfate solution.

Principle

Brass is an alloy of copper and zinc. When dissolved in conc. nitric acid, both the metals get converted into their nitrates.

$$Cu + 4HNO_3 \rightarrow Cu(NO_3)_2 + 2H_2O + 2NO_2 \uparrow \text{ (Brown fumes)}$$

When brass solution is treated with excess of potassium iodide solution, cupric ions oxidize potassium iodide to liberate equivalent quantity of iodine.

$$2Cu(NO_3)_2 + 4KI \rightarrow Cu_2I_2 \downarrow + 4KNO_3 + I_2$$

The liberated iodine is titrated with sodium thiosulfate solution using starch as indicator.

$$NaS_2O_3 + I_2 \rightarrow Na_2S_4O_6 + 2NaI$$

(Sodium tetrathionate)

The oxides of nitrogen present in the brass solution are destroyed by adding urea. The presence of nitrogen oxides will be responsible for the liberation of extra iodine from KI, as they are also good oxidizing agents.

The nitric acid present in the brass solution is neutralized by adding NH_4OH solution till a pale blue precipitate of $Cu(OH)_2$ is obtained. Otherwise, being an oxidizing agent, nitric acid may also liberate iodine from KI. The $Cu(OH)_2$ precipitate is dissolved with dilute acetic acid. Other mineral acids are not preferable as they will bring down the pH to a very small value, at which the liberation of I_2 from KI by Cu^{2+} is not quantitative.

Starch reacts with iodine to form a blue colored complex. At the end point, when free iodine is exhausted in the solution, added quantity of sodium thiosulfate dissociates the starch-iodine complex, consumes iodine and liberates starch, thereby discharging the blue color. For brass solution, the color change is from blue to white as CuI_2 precipitate is present in the solution.

Starch indicator is added near the end point and not in the beginning for the reasons that (i) at high concentration of I_2, starch may form an insoluble starch-I_2 complex and (ii) the iodine also gets adsorbed on colloidal starch particles in the starch solution. The adsorbed iodine cannot be liberated by the addition of sodium thiosulfate, and the amount of iodine getting adsorbed is more when iodine concentration is high. Thus the error is minimized by adding starch near the end point.

During the liberation of I_2 from KI, Cu^{2+} gets reduced to Cu^+ with a change in oxidation state by 1.

Thus equivalent weight of Cu = atomic weight = 63.54

Procedure

Estimation of copper in brass: **Weigh accurately about 1 g of brass alloy. Transfer into a clean beaker. Dissolve the alloy in 1/4 t.t. of conc. nitric acid. Then add about a gram of urea and 1 t.t. of distilled water; boil for 1–2 minutes. Cool the product and transfer quantitatively into a 250ml standard flask. Make up the solution to the mark and shake well for uniform concentration.**

Pipette out 25 ml of brass solution into a clean conical flask. Add NH_4OH solution drop wise until a bluish white precipitate persists. Add dilute acetic acid drop wise just to dissolve the precipitate and then add about 5 ml in excess, followed by the addition of l0ml of 10% KI solution. Titrate the liberated iodine against sodium thiosulfate solution taken in the burette until the color of the solution changes to pale yellow. At this stage add starch solution (approx. 1 ml) as indicator and continue the titration till the blue color changes to white. Repeat the titration for concordant values.

Result

Percentage of copper in the given sample of brass = ...

Observations and Calculation

Estimation of copper in brass

Weight of w. bottle + brass, W_1 =

Weight of w, bottle alone, W_2 =

Weight of brass........ W =

In burette : $Na_2S_2O_3$ solution

In conical flask : 25 ml Brass solution + NH_4OH (till bluish white ppt) + Acetic acid to dissolve the ppt, and 5ml in excess + l0ml 10% KI solution

Indicator : Starch solution (approx.1 ml) near the end point

Colour change : Blue to white.

Burette level	*I*	*II*	*III*
Final			
Initial			
Vol of $Na_2S_2O_3$, ml			

Agreeing Value, V_1 = ... ml

V_1 ml of $Na_2S_2O_3$ solution reacts completely with iodine liberated by Cu^{2+} present in 25 ml of brass solution.

Normality of $Na_2S_2O_3$ solution, $N_1 =$

Normality of Cu^{2+} solution, $N_2 = \frac{V_1 \times N_1}{25} = \ldots$

$= \ldots\ldots\ldots\ldots\ldots$

Weight of copper in a liter of the solution, W_3 = Normality of Cu^{2+} × Eq. wt of Cu

$= N_2 \times 63.54$

$= \ldots$

$= \ldots$ g

Weight of copper in 250 ml of the solution, $W_4 = \frac{W_3}{4}$

$= \ldots$

$= \ldots$ g

Percentage of copper in brass $= \frac{W_4 \times 100}{\text{wt of brass}}$

$= \frac{W_4 \times 100}{W}$

$= \ldots$

$= \ldots$

Signature of the staff member

4. ESTIMATION OF IRON IN HAEMATITE

Aim

To estimate the percentage of iron in haematite, being supplied with a solution containing 10 g of haematite/litre.

Principle

Haematite is an oxide ore of iron, existing as ferric oxide (Fe_2O_3). On treating with HCl, ferric oxide dissolves in the acid as ferric chloride.

$$Fe_2O_3 + 6HCl \rightarrow 2FeCl_3 + 3H_2O$$

Since iron in $FeCl_3$ is already in the highest possible stable oxidation state (Fe^{3+}), it can not be oxidized further. Therefore, it is first reduced to ferrous state and then titrated with potassium dichromate solution, which is an oxidizing agent. Ferric chloride is reduced by stannous chloride in presence of hydrochloric acid.

$$2FeCl_3 + SnCl_2 \rightarrow 2FeCl_2 + SnCl_4$$

$$K_2Cr_2O_7 + 8HCl \rightarrow 2KCl + 2CrCl_3 + 4H_2O + 3(O)$$

$$[2FeCl_2 + 2HCl + (O) \rightarrow 2FeCl_3 + H_2O] \times 3$$

$$K_2Cr_2O_7 + 14HCl + 6FeCl_2 \rightarrow 2KCl + 2CrCl_3 + 6FeCl_3 + 7H_2O$$

From the equation, 2 moles of $FeCl_2$ == 2 moles of $FeCl_3$ == 1 atom of oxygen

∴ Equivalent weight of iron = atomic weight of Fe = 55.85

A slight excess of stannous chloride is added to ensure the complete reduction of $FeCl_3$ to $FeCl_2$. The excess of $SnCl_2$ added is destroyed by adding strong solution of mercuric chloride, since $SnCl_2$ can reduce potassium dichromate.

$$SnCl_2 + 2HgCl_2 \rightarrow SnCl_4 + Hg_2Cl_2 \downarrow \text{ (silky white)}$$

A silky white precipitate of mercurous chloride should be obtained. A black precipitate of finely divided mercury may be produced if large excess of $SnCl_2$ is present or if mercuric chloride solution is added slowly (local excess of $SnCl_2$ is formed). The finely divided mercury reduces $K_2Cr_2O_7$ and also slowly reduces ferric ions formed to ferrous ions. Therefore, the trial must be discarded if a black precipitate is produced. The trial should also be discarded if no precipitate is obtained on the addition of mercuric chloride because the complete reduction of ferric ions to ferrous ions is not ensured.

Potassium ferricyanide is used as an external indicator to detect the end point of the titration. Potassium ferricyanide reacts with ferrous ions to produce an intense deep color due to the formation of ferrous ferric cyanide complex.

$$2K_3[Fe(CN)_6] + 3Fe^{2+} \rightarrow \underset{\text{Ferrous ferricyanide (Blue)}}{Fe_3[Fe(CN)_6]_2} + 6K^+$$

Therefore when a drop of titrated solution is treated with a drop of potassium ferricyanide solution, blue color is developed as long as ferrous ions are present in the solution. At the end point when ferrous ions are completely converted into ferric ions, a drop of the titrated solution does not develop blue color with the indicator.

A standard solution of $K_2Cr_2O_7$ is prepared by dissolving a known weight of the salt in a known volume of the solution.

Procedure

Preparation of standard $K_2Cr_2O_7$ solution: Weigh accurately about 1.25g of $K_2Cr_2O_7$ crystals into a 250 ml standard flask. Dissolve the salt in little distilled water. Make up the solution to the mark and shake well for uniform concentration.

Estimation of iron: Pipette out 25ml of haematite solution into a clean conical flask. Add about 1/4 test tube of conc. HCl and heat to boiling. To the hot solution, add $SnCl_2$ solution from a burette till the yellow color is discharged completely. Add two drops in excess. Cool the product and add 5ml of $HgCl_2$ solution at a stretch and shake well. A silky white precipitate should be obtained. (If black precipitate or no precipitate is obtained, discard the product). Add two test tubes of distilled water.

Place a number of drops of freshly prepared potassium ferricyanide solution on a wax paper. Add a small quantity of potassium dichromatic solution from the burette into the conical flask and shake well. With the help of a thin glass rod take a drop of the titrated solution and bring it in contact with the indicator drop. The color of the indicator drop turns blue. Now add l ml of $K_2Cr_2O_7$ solution at a time into the conical flask and after each addition of l ml $K_2Cr_2O_7$ solution, shake the solution well and test the solution drops as before. Repeat this till a drop of the titrated solution does not develop blue color with the indicator. This titration gives the range of the end point.

Again pipette out 25 ml of the haematite solution into a conical flask. Treat with $SnCl_2$ and $HgCl_2$ as before. Add 2 t.t. disstilled water. This time add most of the potassium dichromate solution required at a stretch and further titrate by the drop wise addition of potassium dichromate solution. After the addition of each drop, test a drop of the titrated solution with the indicator solution drops as explained earlier, till no blue color is developed with the indicator. Repeat for agreeing values.

Result

Percentage of iron in the given haematite sample ....

Observations and Calculations

Weight of haematite ore in a dm^3 of the solution, W = g

Preparation of standard $K_2Cr_2O_7$ solution

Weight of w. bottle + $K_2Cr_2O_7$, W_1 =
Weight of w. bottle alone, W_2 =
Weight of $K_2Cr_2O_7$, W_3 =

Normality of $K_2Cr_2O_7$ solution, $N_1 = \dfrac{\text{Wt. of } K_2Cr_2O_7}{\text{Eq. wt}} \times 4$

$$= \frac{W_3 \times 4}{49.03}$$

= ...

Estimation of iron

In burette : $K_2Cr_2O_7$ solution
In conical flask : 25 ml of haematite solution + 1/4 tt conc. HCl (heat to boiling) + $SnCl_2$ solution, cool, + 5 ml of $HgCl_2$ solution + 2. t.t. of water.

Indicator : Potassium ferricyanide (External indicator)
Colour change : Blue to color less

Burette level	*I*	*II*	*III*
Final			
Initial			
Vol of $K_2Cr_2O_7$, ml			

Agreeing value, V_1 =... ml

V_1 ml of $K_2Cr_2O_7$ solution of...N oxidizes 25 ml of reduced haematite solution completely.

Normality of haematite solution $N_2 = \dfrac{V_1 \times N_1}{25}$

=

= ...

Weight of iron per liter of the solution, $W_4 = N_2 \times$ eq. wt. of iron

$= \ldots \times 55.85$

$= \ldots$

Percentage of iron in the given haematite sample $= \dfrac{W_4 \times 100}{W}$

$=$

$= \ldots$

Signature of the staff member

5. DETERMINATION COD OF WASTE WATER

Aim

To estimate the chemical oxygen demand of waste water.

Principle

Chemical oxygen demand (COD) parameter has been introduced to measure the total oxidisable impurities present in the sewage. This includes both, biologically oxidisable and biologically inert but chemically oxidisable impurities in the water. It is the amount of oxygen equivalent used while oxidizing the chemically oxidisable impurities present in water, with a strong chemical oxidant, $K_2Cr_2O_7$ in acid medium. It is expressed in $mg.dm^{-3}$ or ppm.

A known volume of the waste water sample is refluxed with excess of $K_2Cr_2O_7$ solution in sulfuric acid medium and in the presencc ff Ag_2SO_4 and $HgSO_4$. $K_2Cr_2O_7$ oxidizes all oxidisable impurities. Ag_2SO_4 catalyses the oxidation of straight chain organic compounds, aromatics and pyridine. $HgSO_4$ avoids the interference of Cl^- ions by forming soluble complex with them. In the absence of $HgSO_4$, Cl^- ions precipitate silver ions as AgCl. The amount of unreacted $K_2Cr_2O_7$ is determined by titration with standard Mohr's salt (ferrous ammonium sulfate) solution. The amount of $K_2Cr_2O_7$ solution consumed corresponds to the COD of the water sample. To calculate the COD, a blank titration without the water sample is carried out.

Ferroin (1, 10-phenanthraline-iron (II) complex) is used as the indicator in the titration to detect the end point. The indicator is intensely red in color. Strong oxidizing agents oxidize the indicator to iron (III) complex, which has a pale blue color.

$$\underset{\text{Pale blue}}{[Fe(C_{12}H_8N_2)_3]^{3+}} + e^- \leftrightarrow \underset{\text{Deep red}}{[Fe(C_{12}H_8N_2)_3]^{2+}}$$

As long as $K_2Cr_2O_7$ is present in the solution, the indicator is in the oxidized form, imparting blue color to the solution. The solution becomes blue green as the titration continues due to the formation of Cr(III) ions as a result of reduction of $Cr_2O_7^{2-}$ ions. When $K_2Cr_2O_7$ is completely exhausted in the solution, the oxidized form of the indicator gets reduced to the reduced farm, imparting red-brown color to the solution. Therefore, the end paint is marked by the color change from blue-green to red-brown. The end point is sharp only at high acid concentration and therefore, the solution is kept at a very high acidity of H_2SO_4.

A standard solution of Mohr's salt is prepared by dissolving a known weight of the salt in a known volume of the solution. A test tube of dil. H_2SO_4 acid is added during dissolution to prevent the hydrolysis of the salt in the solution.

The reaction between Mohr's salt and $K_2Cr_2O_7$ can be represented as follows.

$$K_2Cr_2O_7 + 4H_2SO_4 \rightarrow K_2SO_4 + Cr_2(SO_4)_3 + 4H_2O + 3(O)$$

$$2FeSO_4 + H_2SO_4 + (O) \rightarrow Fe_2(SO_4)_3 + H_2O] \times 3$$

$$K_2Cr_2O_7 + 6FeSO_4 + 7H_2SO_4 \rightarrow K_2SO_4 + Cr_2(SO_4)_3 + 3Fe(SO_4)_3 + 7H_2O$$

Equivalent weight of Mohr's salt = Molecular weight = 392

Procedure

Preparation of standard Mohr's salt solution: Weigh accurately about 9 g of Mohr's salt, transfer into a 250 ml standard flask, dissolve in a little distilled water, add a test tube of dilute H_2SO_4 and make up to the mark with distilled water. Shake well for uniform concentration.

Determination of COD: Pipette out 25ml of the waste water sample into a 250 ml conical flask with a ground glass neck. Add 1 g of $HgSO_4$, followed by 2 test tubes of silver sulfate / sulfuric acid solution (prepared by dissolving about 5 g of Ag_2SO_4 in 250 ml of conc. Sulfuric acid and then slowly diluting to 500 ml with distilled water). Then pipette out 25ml of $K_2Cr_2O_7$ solution into the flask. Fit the flask with the reflex condenser and reflux the mixture for about half an hour. On cooling, rinse the inside of the condenser with distilled water into the flask. Add 2–3 drops of ferroin indicator and titrate with standard Mohr's salt solution till the color changes from blue green to reddish brown.

Blank Titration: Repeat the above procedure by taking 25 ml distilled water in the place of sample waste water.

Result

COD of the given sample of water = ...

Observations and Calculation

Preparation of standard Mohr's salt solution;

Weight of w. bottle + Mohr's salt, $W_1 =$

Weight of w. bottle alone, $W_2 =$

Weight of Mohr's salt, $(W_1 - W_2) = W =$

Normality of Mohr's salt solution, $N_1 = \dfrac{W \times 4}{\text{Eq. wt. of Mohr'salt}}$

$$= \frac{W \times 4}{392}$$

$=$

$= \ldots$

Volume of Mohr's salt consumed by excess of $K_2Cr_2O_7$ after oxidizing impurities in sample water = $V_1 = \ldots$ ml

Volume of Mohr's salt solution consumed in blank titration $= V_2 = \ldots$ ml

25ml of $K_2Cr_2O_7$ solution + 25ml of water sample $\equiv V_1$ ml of N_1 N Mohr's salt solution.

25mlof $K_2Cr_2O_7$ solution alone $\equiv V_2$ ml of N_1N Mohr's salt solution.

Volume of $K_2Cr_2O_7$ solution consumed by oxidisable impurities in terms of volume of Mohr's salt solution $= (V_2 - V_1)$ ml $=$

$= \ldots$

1000 ml of 1 N Mohr's salt solution $=$ 8 g of oxygen

1 ml of 1N Mohr's salt solution $= \frac{8}{1000}$ g of oxygen

$(V_2 - V_1)$ml of N_1 N Mohr's salt solution, $W_3 = \frac{8(V_2 - V_1) \times N_1}{1000}$ g of oxygen

$=$ wt. of COD in 25ml of waste water.

$= \ldots$ g

$= \ldots$ mg

COD of the sample $= \frac{W_3}{25} \times 1000$

$=$

$= \ldots$ mg. dm^{-3}

Signature of the staff member.

6. DETERMINATION OF TOTAL ALKALINITY OF WATER

Aim:

To determine the total alkalinity of a given sample of water, by titration with standard HCl solution.

Principle

The alkalinity of water is due to the presence of hydroxyl ions (OH^-) from NaOH or KOH, carbonate ions (CO_3^{2-}) from sodium carbonate or potassium carbonate and bicarbonates (HCO_3^-) from sodium bicarbonate or potassium bicarbonate or temporary hardness (calcium bicarbonate and magnesium bicarbonate).

The alkalinity of water can be estimated by titrating water with a standard acid using phenolphthalein and methyl orange as indicators.

When titrated with water, the reactions taking place are as follows:

i) $OH^- + H^+ \rightarrow H_2O$

ii) $CO_3^{2-} + H^+ \rightarrow HCO_3^-$

iii) $HCO_3^- + H^+ \rightarrow H_2O + CO_2$

Titration of the water sample with a standard acid to methyl orange end point marks the completion of all the three reactions above. Hence, the amount of the total acid consumed correspond to the total alkalinity of the water sample.

Procedure

Pipette out 50 ml of the water sample into a conical flask. Add 2 – 3 drops of methyl orange indicator and titrate with the standard acid solution till colour changes from yellow to pink. Note down the titre value. Repeat for agreeing values.

Result:

Total alkalinity of the given sample of water =............... Ppm

Observations and Calculations

Volume of water sample taken $= V_1 = 50$ ml

Concentration of the standard hydrochloric acid solution $= N_1 =$ ....N

Volume of acid consumed $= V_2 =$ ....ml

Concentration of total alkalinity $= N_2 = \frac{V_2 N_1}{50}$

$= \ldots\ldots\ldots N$

Total alkalinity in terms $CaCO_3$ equivalents $= N_2 \times 50$ g/litre

$= \ldots$ g/litre $= \frac{N_2 \times 50 \times 10^6}{100}$

$= \ldots$ ppm

Signature of the staff member

Index